AF368912

Down Syndrome: Neuropsychological Phenotype across the Lifespan

Down Syndrome: Neuropsychological Phenotype across the Lifespan

Editor

Margaret B. Pulsifer

MDPI • Basel • Beijing • Wuhan • Barcelona • Belgrade • Manchester • Tokyo • Cluj • Tianjin

Editor
Margaret B. Pulsifer
Department of Psychiatry,
Massachusetts General Hospital
Harvard Medical School
Boston
United States

Editorial Office
MDPI
St. Alban-Anlage 66
4052 Basel, Switzerland

This is a reprint of articles from the Special Issue published online in the open access journal *Brain Sciences* (ISSN 2076-3425) (available at: www.mdpi.com/journal/brainsci/special_issues/ Neuropsychological_Down).

For citation purposes, cite each article independently as indicated on the article page online and as indicated below:

LastName, A.A.; LastName, B.B.; LastName, C.C. Article Title. *Journal Name* **Year**, *Volume Number*, Page Range.

ISBN 978-3-0365-3962-1 (Hbk)
ISBN 978-3-0365-3961-4 (PDF)

Contents

brain sciences

Editorial

Down Syndrome: Neuropsychological Phenotype across the Lifespan

Margaret Pulsifer

Department of Psychiatry, Massachusetts General Hospital, Harvard Medical School, Boston, MA 02114, USA; mpulsifer@mgh.harvard.edu

Citation: Pulsifer, M. Down Syndrome: Neuropsychological Phenotype across the Lifespan. *Brain Sci.* **2021**, *11*, 1380. https://doi.org/10.3390/brainsci11111380

Received: 8 October 2021
Accepted: 18 October 2021
Published: 21 October 2021

Publisher's Note: MDPI stays neutral with regard to jurisdictional claims in published maps and institutional affiliations.

Down syndrome (DS), caused by triplication of chromosome 21, is the most common genetic cause of intellectual disability (ID), with an estimated incidence of one in 700 live births. Individuals with DS commonly exhibit unique neuropsychological profiles that emerge during specific developmental stages across the lifespan, often characterized by early developmental delay, cognitive strengths and weaknesses, behavior and mental health issues and age-related cognitive decline frequently resulting in early onset Alzheimer's disease. These profiles are unique compared to other individuals with ID and reflect the genetic mechanisms and neuroanatomic features underlying the distinct neuropsychological phenotype associated with DS. Understanding this neuropsychological phenotype across the lifespan and the associated clinical, educational and treatment needs is particularly important because of recent increases in life expectancy due to improved health care, advocacy, services and societal changes.

Individuals with Down syndrome often face unique developmental, cognitive and behavioral challenges at various stages of development across the lifespan. This Special Issue contains original research and comprehensive reviews that address a broad range of topics related to DS, including early developmental trajectories, social challenges and autism spectrum disorder, acute developmental regression, psychiatric issues and treatment, assessment and diagnosis, early onset dementia and medical co-morbidities.

Communication skills in infants with DS are often delayed compared to typically developing infants, but little is known about what factors are associated with this developmental delay. In a unique study, Pejovic et al. [1] compared 5–7-month-old infants with DS to typically developing infants in terms of visual attention and audiovisual speech processing and found that the infants with DS were slower to orient visual attention and less attentive to social/speech cues. Those differences might be related to delays in subsequent communication development. The findings could help direct more effective interventions, such as providing more time for infants with DS to attend to communicative cues and promoting attention to cues by emphasizing face-to-face communication.

Examining speech fluency, Naess et al. [2] found significantly greater difficulties with fluency in a group of children with DS with an average age of 6 years compared to typically developing children with a similar level of nonverbal cognitive functioning. Poorer performance in speech fluency was associated in the study with lower language skills. The findings highlight the need for interventions that target both speech fluency and language development in children with DS.

Another study by Naess et al. [3] examined reading skills in a national cohort of school aged children with DS in Norway, age 6 to 8 years. Children with DS often experience reading difficulties; the study's goal was to identify factors that were associated with better decoding skills by 3rd grade. The most important factors, after controlling for nonverbal cognitive abilities, were vocabulary and letter knowledge. Those findings should help guide early educational interventions.

Autism spectrum disorder (ASD) is more commonly diagnosed in individuals with DS than in the general population. However, little is known about the unique presentation of ASD symptoms in DS. In a study examining the prevalence and presentation of ASD in a large cohort of DS individuals, age 6 to 23 years, Dimachkie Nunnally et al. [4] found that 37% of the cohort were classified as ASD on a semi-structured standardized assessment. That ASD group was marked by greater challenges with social communication and complex expressive language than those without ASD.

Children with DS, with and without ASD, experience social challenges which may result in social isolation and may impact mental health outcomes. Therefore, accurate measures of social cognition and social behavior are particularly important for children and adolescents with DS. Schworer et al. [5] addressed this need by examining the psychometric properties of several measures in a group of individuals with DS, age 6 to 17 years. One measure, the Social Responsiveness Scale-2, was particularly promising for this population.

Levels of daily living skills and independence vary in individuals with DS across the lifespan. In a large cohort of pediatric and adult individuals with DS, Krell et al. [6] found that 87% communicated verbally and fewer (17%) could use written communication. The life skills reported as most important to both adolescents and adults included learning about healthy food, preparing meals and describing symptoms to a doctor. The authors recommended that life skills should be routinely assessed during a medical appointment to support greater independence for individuals with DS.

Between teenage years and the mid to late 20 s, some individuals with DS experience an acute regression in skills and behavior. There is limited understanding of the cause of this unexplained condition and its diagnosis, treatment and prognosis. Two articles in this Special Issue examine regression in DS; one, by Walpert et al. [7], presents a review of 13 articles on early regression in DS and describes symptom presentation, potential trigger events and prognosis. The other, by Handen et al. [8], uses biomarker risk measures for Alzheimer's disease (AD) to examine the possibility that early regression in adults with DS might lead to increased risk of AD in later life. The findings have important implications for recognition of the condition and highlight the need for further research to focus on prevention and treatment.

Depression is common in individuals with DS, but treatment with serotonin reuptake inhibitors (SSRIs) has not been closely examined in this population. In a retrospective chart review, Thom et al. [9] evaluated the effectiveness, tolerability and safety of SSRIs for depression in a cohort of adults with DS. The majority of patients in the study responded to a 12-week course of SSRI treatment, with some tolerating long-term use.

Adults with DS have an especially high risk for developing early onset AD. Accurate identification and assessment of cognitive decline in this population are essential for diagnosis and planning. Two articles in this Special Issue, by Hom et al. and Harp et al., examine the effectiveness of multiple measures of cognitive and behavioral functioning and the domains assessed in identifying dementia classification. The research by Hom et al. [10] shows that cognitive functioning can be characterized at the cognitive domain level, with language, executive functioning and memory being candidates for most impacted domains. Building on this concept, Harp et al. [11] developed an abbreviated test battery to identify individuals with DS at risk for AD.

AD and hypothyroidism are equally prevalent in adults with DS, but the relationship between the age of onset of hypothyroidism and that of AD has not been established. In a study that is the first to explore this relationship, Lai et al. [12] suggest that early onset hypothyroidism in DS is significantly associated with an early onset age of AD, independent of several co-occurring conditions of interest, including APOE ε4 allele status. The authors recommend early testing of thyroid functions in adults with DS and emphasize the need for future studies to determine how hypothyroidism affects AD risk and onset.

Thanks are due to all of the authors contributing to this Special Issue. Their work represents meaningful advances in the understanding of Down syndrome throughout the lifespan, with benefits both for research and for clinicians working to improve the well-being of individuals with Down syndrome.

Funding: Within the past three years, Dr. Pulsifer has received funding from the National Institutes of Health (U01 AG051412 (ADDS) and U01 DC019279) and the LuMind IDSC Down Syndrome Foundation for research for individuals with Down syndrome.

Conflicts of Interest: Margaret Pulsifer declares no conflict of interest.

References

1. Pejovic, J.; Cruz, M.; Severino, C.; Frota, S. Early Visual Attention Abilities and Audiovisual Speech Processing in 5–7 Month-Old Down Syndrome and Typically Developing Infants. *Brain Sci.* **2021**, *11*, 939. [CrossRef] [PubMed]
2. Næss, K.-A.B.; Nygaard, E.; Hofslundsengen, H.; Yaruss, J.S. The Association between Difficulties with Speech Fluency and Language Skills in a National Age Cohort of Children with Down Syndrome. *Brain Sci.* **2021**, *11*, 704. [CrossRef] [PubMed]
3. Næss, K.-A.B.; Nygaard, E.; Smith, E. Occurrence of Reading Skills in a National Age Cohort of Norwegian Children with Down Syndrome: What Characterizes Those Who Develop Early Reading Skills? *Brain Sci.* **2021**, *11*, 527. [CrossRef] [PubMed]
4. Dimachkie Nunnally, A.; Nguyen, V.; Anglo, C.; Sterling, A.; Edgin, J.; Sherman, S.; Berry-Kravis, E.; del Hoyo Soriano, L.; Abbeduto, L.; Thurman, A.J. Symptoms of Autism Spectrum Disorder in Individuals with Down Syndrome. *Brain Sci.* **2021**, *11*, 1278. [CrossRef]
5. Schworer, E.K.; Hoffman, E.K.; Esbensen, A.J. Psychometric Evaluation of Social Cognition and Behavior Measures in Children and Adolescents with Down Syndrome. *Brain Sci.* **2021**, *11*, 836. [CrossRef]
6. Krell, K.; Haugen, K.; Torres, A.; Santoro, S.L. Description of Daily Living Skills and Independence: A Cohort from a Multidisciplinary Down Syndrome Clinic. *Brain Sci.* **2021**, *11*, 1012. [CrossRef] [PubMed]
7. Walpert, M.; Zaman, S.; Holland, A. A Systematic Review of Unexplained Early Regression in Adolescents and Adults with Down Syndrome. *Brain Sci.* **2021**, *11*, 1197. [CrossRef] [PubMed]
8. Handen, B.; Clare, I.; Laymon, C.; Petersen, M.; Zaman, S.; O'Bryant, S.; Minhas, D.; Tudorascu, D.; Brown, S.; Christian, B.; et al. Acute Regression in Down Syndrome. *Brain Sci.* **2021**, *11*, 1109. [CrossRef] [PubMed]
9. Thom, R.P.; Palumbo, M.L.; Thompson, C.; McDougle, C.J.; Ravichandran, C.T. Selective Serotonin Reuptake Inhibitors for the Treatment of Depression in Adults with Down Syndrome: A Preliminary Retrospective Chart Review Study. *Brain Sci.* **2021**, *11*, 1216. [CrossRef] [PubMed]
10. Hom, C.L.; Kirby, K.A.; Ricks-Oddie, J.; Keator, D.B.; Krinsky-McHale, S.J.; Pulsifer, M.B.; Rosas, H.D.; Lai, F.; Schupf, N.; Lott, I.T.; et al. Cognitive Function during the Prodromal Stage of Alzheimer's Disease in Down Syndrome: Comparing Models. *Brain Sci.* **2021**, *11*, 1220. [CrossRef] [PubMed]
11. Harp, J.P.; Koehl, L.M.; Pelt, K.L.V.; Hom, C.L.; Doran, E.; Head, E.; Lott, I.T.; Schmitt, F.A. Cognitive and Behavioral Domains That Reliably Differentiate Normal Aging and Dementia in Down Syndrome. *Brain Sci.* **2021**, *11*, 1128. [CrossRef] [PubMed]
12. Lai, F.; Mercaldo, N.D.; Wang, C.M.; Hersch, M.S.; Hersch, G.G.; Rosas, H.D. Association between Hypothyroidism Onset and Alzheimer Disease Onset in Adults with Down Syndrome. *Brain Sci.* **2021**, *11*, 1223. [CrossRef] [PubMed]

Article

Early Visual Attention Abilities and Audiovisual Speech Processing in 5–7 Month-Old Down Syndrome and Typically Developing Infants

Jovana Pejovic *, Marisa Cruz, Cátia Severino and Sónia Frota

Center of Linguistics, Lisbon Baby Lab, University of Lisbon, 1600-214 Lisbon, Portugal;
marisac@edu.ulisboa.pt (M.C.); catiaseverino@campus.ul.pt (C.S.); sfrota@edu.ulisboa.pt (S.F.)
* Correspondence: jpejovic@edu.ulisboa.pt

Abstract: Communicative abilities in infants with Down syndrome (DS) are delayed in comparison to typically developing (TD) infants, possibly affecting language development in DS. Little is known about what abilities might underlie poor communication and language skills in DS, such as visual attention and audiovisual speech processing. This study compares DS and TD infants between 5–7 months of age in a visual orientation task, and an audiovisual speech processing task, which assessed infants' looking pattern to communicative cues (i.e., face, eyes, mouth, and waving arm). Concurrent communicative abilities were also assessed via the CSBS-DP checklist. We observed that DS infants orient their visual attention slower than TD infants. Both groups attended more to the eyes than the mouth, and more to the face than the waving arm. However, DS infants attended less to the eyes than the background, and equally to the face and the background, suggesting their difficulty to assess linguistically relevant cues. Finally, communicative skills were related to attention to the eyes in TD, but not in DS infants. Our study showed that early attentional and audiovisual abilities are impaired in DS infants, and might underlie their communication skills, suggesting that early interventions in this population should emphasize those skills.

Keywords: attention; audiovisual processing; Down syndrome; communicative abilities in infants

Citation: Pejovic, J.; Cruz, M.; Severino, C.; Frota, S. Early Visual Attention Abilities and Audiovisual Speech Processing in 5–7 Month-Old Down Syndrome and Typically Developing Infants. *Brain Sci.* **2021**, *11*, 939. https://doi.org/10.3390/brainsci11070939

Academic Editor: Margaret B. Pulsifer

Received: 2 June 2021
Accepted: 13 July 2021
Published: 16 July 2021

Publisher's Note: MDPI stays neutral with regard to jurisdictional claims in published maps and institutional affiliations.

1. Introduction

Down syndrome (DS) is associated to a genetic perturbation known as trisomy 21 affecting physical, motor, and cognitive functioning. It is the most common genetic cause of intellectual disability. DS vastly affects language processing and development [1–4]. Both language comprehension and production deficits have been described. In particular, growth slopes in comprehension become shallower with age and language production studies demonstrate either delayed or atypical speech patterns (for a review, see [5]), especially in childhood and adolescence (e.g., [6]). Speech production in infants and toddlers revealed mixed results (e.g., [7,8]). Phonological acquisition has been reported to be delayed, showing deviant patterns [9], and comprehension and production of prosody has been shown to be impaired in children with DS and adolescents [10]. Equally important, hearing in individuals with DS is often impaired consequently affecting their language learning (for a review, see [3,11]). Language learning difficulties are evident in the late occurrence of first words/signs which appear between 24–36 months of age, while in typically developing children they usually occur between 12 and 18 months of age. In addition to language processing, DS 26-month-old toddlers show impairment in their social communication abilities (e.g., [12]). Taken together, speech impairment in DS extends to later developmental stages affecting DS individuals' overall communicative skills, and possibly academic success, and general well-being.

Many studies identified the benefit of early parent-implemented intervention in DS children younger than three years of age for their further language skills (for a review, see, e.g., [12]). One of the main aims of these early interventions in DS children is to target abilities that might relate to later language outcomes. For instance, a recent meta-study on joint attention demonstrated that this ability is rather a strength than a weakness in DS population [13]. Developmentally, joint attention refers to a nonverbal skill occurring in social interaction between an infant and a caregiver. Using eye-gaze cues, pointing gestures or vocalizations, attention between the infant and the caregiver is focused/shared to the same object/event, and accompanied by awareness that the attentional focus is shared (e.g., [14]). Joint attention in typically developing populations is related to object learning [15], word learning (e.g., [16,17]; but see other proposals [18,19]), or later language outcomes (e.g., [20]). Similarly, joint attention is relevant for word learning in Down syndrome children as well [21], and it is a strong predictor in DS infants for later expressive and receptive language outcomes [12]. However, this skill emerges chronologically later in DS children (in their second year of life) than in typically developing children (for a review, see, e.g., [22]), suggesting that precursors to joint attention development might be impaired in Down syndrome children. In the present study, we will focus on some of these possible cognitive abilities that might directly or indirectly support further language development in DS population.

To support initial communicative abilities, infants have to learn to take part in non-verbal communication (i.e., joint attention), but they also have to selectively attend to relevant social communicative cues. In particular, they need to attend to faces and communicative gestures, and to process visual communicative cues accompanying the auditory speech signal (i.e., articulatory movements, eyebrows and head movements, gestures, etc.). In adults, attending to these cues facilitates face-to-face communication in noisy conditions (e.g., [23,24]). In infants, visual cues may support phonetic and word learning [25,26], as well as the learning of syntax [27]. Importantly, visual cues are available to infants already in early infancy—by four months of age infants are able to integrate auditory and visual information (e.g., [28–30]). Thus, the ability to attend to visual communicative cues develops early in infancy and is important for language development. Understanding infants' attention to visual communicative cues in atypically developing populations is particularly relevant, especially for infants undergoing speech interventions that are often based on improving communicative abilities.

Studies investigating the ability to process visual communicative cues suggest that DS infants are delayed in comparison to chronologically matched TD infants. For instance, DS infants discriminate between objects and human faces by four months of age, while TD infants do so already by two months of age [31], suggesting impaired ability to detect relevant social communicative cues in DS early development. Further, in a longitudinal study during the first six months of life, TD infants demonstrate a first peak in forming eye contact with their mothers already at one and a halfmonths of age, while DS infants do so around their third month of age [32]. Interestingly, the same study revealed that once DS infants form eye contact, they maintain it longer than the TD group, possibly affecting their ability to shift their gaze towards other objects in their environment that a caregiver is gazing to. A recent study demonstrated that unlike TD toddlers, DS toddlers at 16 months of age (chronologically age matched with a TD group), and at 28 months (mental age matched with the 16-month-old TD group) are not able to detect a mismatch in the audiovisual speech signal [33].

Importantly, attentional (cognitive) impairments in DS infants go beyond the above-mentioned impairments in visual speech processing and attention to faces. DS toddlers are slower in disengaging their visual attention from an object they have been engaged to, in comparison to chronologically or mentally-age matched TD infants, as shown by [34]. The same study showed that being faster in visual attention disengagement relates with higher expressive and receptive vocabulary abilities in both TD and DS toddlers. In another study, five-year-old DS children were faster in disengaging than TD children, but similar in how fast they orient (attend) to visual stimuli [35]. Other study yet reported lower performance in DS children from three-six years of age in visual sustained attention [36]. Therefore, results converge in suggesting that DS children and toddlers' visual attention abilities are impaired in comparison to their TD peers. However, little is known on early visual attention abilities in DS infants, particularly in their first six months of life. Understanding visual attention skills in DS in the first months of life is crucial to understand their reported impairments in early face processing and audiovisual speech processing (e.g., [31,33]), that possibly underlie their impaired language development.

The current study assessed five-to-seven-month-old DS infants and compared them to a chronologically matched TD group in three separate measures of visual attention, audiovisual speech processing, and communication abilities. Our main goal was to establish what the early relations between the three components are and compare them across the DS and TD groups. We hypothesized that DS infants' performance in all the three measures would differ from TD infants' performance. Specifically, we expected that DS infants would show an impairment in visual attention, reflected in slower visual orientation latency, while for the audiovisual task the DS group would attend less to communicative cues than their TD peers. Finally, we expected that the DS group would underperform on measures of communicative abilities in comparison to the TD group.

2. Materials and Methods

2.1. Participants

Seven infants with Down syndrome (mean age = 6 months; age range from 5 months and 3 days to 7 months and 20 days; 3 males) and 24 typically developing infants (mean age = 5.25 months; age range from 5 months and 2 days to 6 months and 28 days; 16 males) took part in this study. DS infants were recruited from the Center for Child Development Diferenças in Lisbon, Portugal. They were born full-term and had normal hearing to mild hearing loss and normal or corrected-to-normal vision (according to clinical screening). TD infants were all born full-term, with no reported medical/developmental concerns. Additionally, questionnaires on language and overall development (see details in Materials and Procedure) served as a screening tool to confirm TD infants' development. All infants were raised in monolingual European Portuguese homes. The study was approved by the Ethical Committee for Research of the School of Arts and Humanities of the University of Lisbon.

2.2. Materials and Procedure

Infants took part in two tasks: (1) the visual attention task, and (2) the audiovisual task. First, infants were tested in the visual attention task, followed by the audiovisual task. After completing both tasks, parents provided information on demographic and health status of the infant. Overall communicative development was assessed with the Communicative and Symbolic Behavior Scales Developmental Profile (CSBS-DP) adapted for Portuguese, that measures infants and toddlers' development from 6 to 24 months of age [37]. The CSBS-DP provides data on several scales: emotion and use of eye gaze, use of communication, use of gestures, use of sounds, use of words, understanding of words, and use of objects.

2.2.1. Visual Orientation Attention Task

Similar to a previous study on DS and TD children [35], we tested infants in a visual orientation attention task. In this task we measured infants' looking latency to visual stimuli, here flashing lights. Infants were seated on their caregiver's lap in a testing booth facing the central green light, while two red lights were placed laterally from the infant. Infants' looking behavior was monitored on a camera (Logitech c920, Logitech, Fremont, CA, USA) and online coded by an experimenter placed outside the booth. Every trial began by flashing the central green light. Once the infant orients toward it, this light turns off and one of the lateral red lights starts to flash. When the infant directs its look toward the lateral red light, the experimenter records this by pushing a button on the keyboard. The lateral red light continues to flash for 2 s and then turns off, while the green-central light starts flashing and a new trial begins. Infants' orientation latency is measured as the time between the onset of flashing of the lateral red light and the moment when infants look away from the central green light towards the lateral red light. If an infant did not direct its look to the central light, the experimenter played a short infant-friendly sound to recover infants' attention to the task. Moreover, if the experimenter noticed that the infant is not attending to the lateral light for a significantly long period (i.e., longer than 8 s), the trial ended by turning off the red light and initiating the green light. The 8-s reference was used as it was the maximum trial duration in [35]. There was a maximum of 10 trials (5 on the left and 5 on the right), and the presentation of the left vs. right lateral trials was randomized. The task stopped if an infant lost interest in the task, therefore the number of trials might vary across infants. Stimuli presentation was controlled by the Look software [38]. The time of infants' orientation latency from the central green light to the lateral red light was also recorded by the software.

2.2.2. Audiovisual Task

Infants' eye-gaze was recorded while watching 4-s-long videos of an animated character (Noddy) talking and waving at the infant (Figure 1). Videos were part of a stress perception task where they were inserted after each block as a reinforcer [39]. Auditorily, four different reinforcing passages were paired with the same video (e.g., "That's it! We are going to play one more time" (Four following passages were used: "É isso! Vamos jogar mais uma vez" (That's it! We are going to play one more time); "Muito bem. Vamos continuar o nosso jogo" (Well done! We are going to continue our game); "Muito bem! Este jogo é muito divertido" (Well done! This game is a lot of fun); "Parabéns! Vejo que estás mesmo a gostar disto" (Good! You are really enjoying the game)). The video did not change visually throughout the task, only the auditory passages. The order of presentation of the video with the different passages was randomized between infants. Intentionally, the Noddy character was presented centrally in the video, against a colorful and attractive background, to assess infants' attention to visual linguistic and paralinguistic communicative cues (the face, the arm) versus non-linguistic objects (the background). In total, infants could be presented with up to eight videos. However, the stress perception task stopped when infants lost interest in the task, therefore infants varied in how many videos they were presented with.

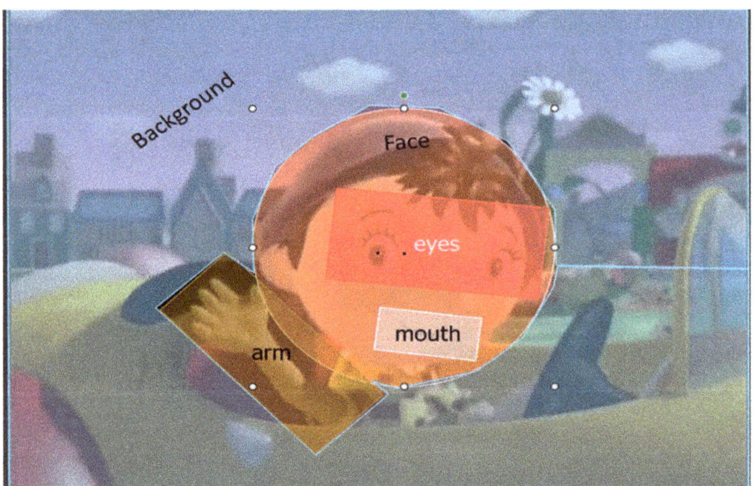

Figure 1. Still-example frame from the audiovisual task. Marked in colors are the areas of interest analyzed in the task: the face, the eyes, the mouth, the arm, and the background.

Infants were seated in the caregiver's lap in a dimmed testing booth, facing the stimuli presentation monitor (Dell LCD screen in 1680 × 1050 pixel resolution) on ~70 cm distance from the monitor. Auditory stimuli were played over speakers (Genious) placed behind the monitor. Infants' eye-gaze was recorded using the SMI RED500 eye-tracker, whereas the SMI Experimenter Center and iView X software-controlled stimuli presentation.

2.2.3. Overall and Communicative Development Assessment

Parents filled in the CSBS-DP checklist at the time of the audiovisual and attention tasks, since this tool was also used as a screening tool to make sure that the TD group indeed followed a typical development. Not all infants that participated in the attention and audiovisual tasks provided data for the questionnaire, and thus the sample of infants for overall and communicative development assessment differed from that of the audio-visual/attention tasks (see details in the result section). Data from the CSBS-DP were examined through correlation analyses with performance on the audiovisual task.

3. Results

3.1. Visual Orientation Task

Infants' latency (in seconds) in orienting to the red lateral light was measured for every trial and averaged for each infant. In both groups, the majority of infants completed all 10 trials. In the DS group, 6 out of 7 infants completed 10 trials (M = 9.28, range from 5–10). In the TD sample, 16 out of 24 infants provided data for all trials (M = 8.95; range from 5–10 trials). Infants' orientation latency for the two groups is provided in Figure 2. Because sample size differed across groups, we performed a linear-mixed model analysis with infants' orientation latency as the dependent variable, group (DS and TD) as a fixed effect, while by-subject intercept was set as a random effect. Using the lmerTest [40] package in R, we observed that the DS group revealed significantly longer latency than the TD group (intercept = 6.29, DS estimate = 2.46, SE = 0.86, t = 2.85, *p* = 0.008, 95% CI: 0.77–4.16). Additionally, we compared the number of trials longer than 8 s across groups. A Wilcoxon-Mann-Whitney test revealed that DS infants exhibited more trials longer than 8 s (M = 3.0, SD = 2.0) than TD infants (M = 1.29, SD = 1.5; Z = 2.19, *p* = 0.028, r = 0.39).

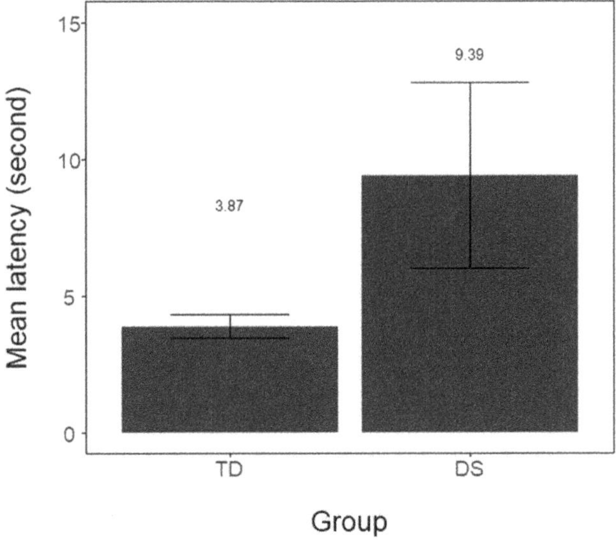

Figure 2. Mean orientation latency (in seconds) in the visual attention task across the Down syndrome (DS) and typically developing (TD) groups. The values above the bars refer to the mean latency value for each group. Error bars represent 1 (+/−) standard error of mean.

3.2. Audiovisual Task

Infants' looking times to the screen were recorded with an eye-tracker. We defined dynamic areas of interest (AOI) covering the background, the arm, the face, the eyes, and the mouth (Figure 1). For each trial, we calculated the proportion of looking time to the AOIs in comparison to the whole screen. Next, for each infant we averaged proportions across all trials. The total number of trials differed across infants, depending on how long they were interested in the task (between 1 and 8 blocks). A Wilcoxon-Mann-Whitney test revealed that groups did not differ in the number of completed trials (M_{TD} = 3.91, range 2–6; M_{DS} = 3.4, range 2–5; $Z = -0.71$, $p = 0.47$, r = 0.13). The looking patterns for the two groups of infants are depicted in Figure 3.

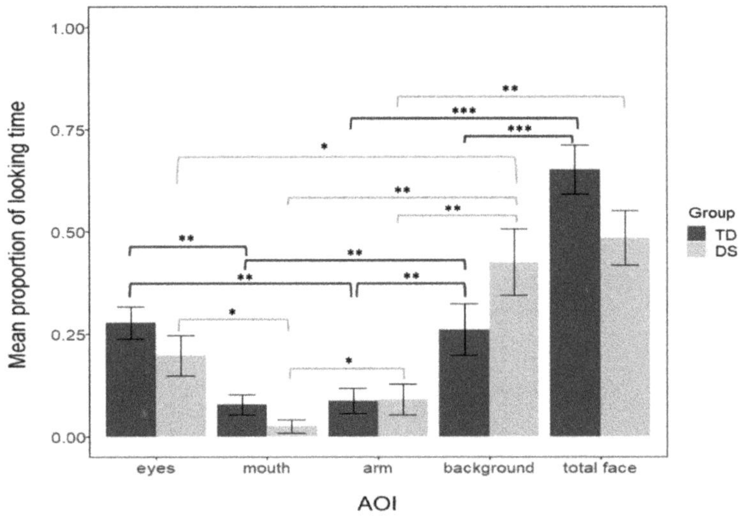

Figure 3. Mean proportion of looking time in the audiovisual task across the Down syndrome (DS) and typically developing (TD) groups. Error bars represent 1 (+/−) standard error of mean. Significant differences are signaled: 0.05 = *, 0.01 = **, 0.001 = ***.

The looking pattern was analyzed for each group separately. First, we compared the 4 levels of AOIs (the eyes, the mouth, the arm, and the background) separately for each group. The results revealed that AOIs significantly differed in both groups (a Kruskal-Wallis test for the DS, H(3) = 17.7, $p = 0.004$, $\eta^2 = 0.61$; a one-way-ANOVA for the TD F(3, 92) = 8.82, $p < 0.001$, $\eta^2 = 0.22$). Pairwise comparisons (Bonferroni controlled) showed that regarding the background, both groups looked longer at the background than the arm (both $ps = 0.001$, d_{DS} = 1.3, d_{TD} = 0.35), and longer to the background than the mouth (p_{TD} = 0.0012, d_{TD} = 1.8, p_{DS} = 0.0017, d_{DS} = 0.66). However, DS infants looked longer at the background than the eyes ($p = 0.04$, d = 0.78), but not TD infants ($p = 0.86$, d = 0.03). Regarding the arm, TD, but not DS infants, looked more at the eyes than the arm (p_{TD} = 0.001, d_{TD} = 0.38; p_{DS} = 0.2, d_{DS} = 0.47), whereas DS, but not the TD, looked more at the arm than the mouth (p_{TD} = 0.86, d_{TD} = 0.01; p_{DS} = 0.039, d_{DS} = 0.88). Finally, both groups looked longer at the eyes than the mouth (p_{TD}= 0.001, d_{TD} = 0.4; p_{DS} = 0.035, d_{DS} = 0.89). To further understand this complex looking pattern across the groups we reduced the number of AOIs, thus we compared the face (including the eyes and the mouth), the arm, and the background separately for each group. We observed that both groups looked more at the background than the arm (p_{TD} = 0.01 d_{TD} = 0.32; p_{DS} = 0.0017, d_{DS} = 1.25) and more at the face than the arm (p_{TD} < 0.001, d_{TD} = 1.02; p_{DS} = 0.0017, d_{DS} = 1.22). However, only TD, but not DS infants, looked more at the face than the background (p_{TD} < 0.001, d_{TD} = 0.71; p_{DS} = 0.6, d_{DS} = 0.18). To directly compare the two groups, we computed a linear-mixed analysis on proportion of looking time with AOI (face, arm, background) and group (TD and DS) as fixed factors (with the interaction term), and by-subject as a random intercept. This analysis confirmed that the AOIs differed (F = 26.94, p < 0.001, η^2_p = 0.38, 95% CI = 0.22–0.51), and more importantly that there was an interaction between AOI and Group (F = 3.21, p = 0.04, η^2_p = 0.07, 95% CI = 0.01–0.18). Further pairwise comparisons (Bonferroni controlled) revealed that groups did not differ in their looking time to the arm (t = −0.03, p = 0.97, d = 0.06). However, results suggest a trend of TD looking more to the face than DS (MTD = 0.65, MDS = 0.48, t = 1.8, p = 0.074, d = 0.38), and a trend for TD looking less to the background than DS (MTD = 0.26, MDS = 0.42, t = −1.77, p = 0.079, d = 0.38).

Finally, considering that DS infants were slower in the visual orientation task, we tested whether individual latency in the visual orientation task modulated performance in the audiovisual task. To the previous mixed model analysis, we added the average latency for each subject as a fixed effect, while other parameters maintained the same. The results were similar as in the previous analysis, with a main effect of AOI (F = 26.64, p < 0.001, η^2_p = 0.38, 95% CI = 0.22–0.51) and an interaction between AOI and Group (F = 3.17, p = 0.04, η^2_p = 0.03, 95% CI = 0.01–0.18). The pairwise comparisons revealed the same pattern: TD and DS do not differ in their proportional looks to the arm (t = −0.02, p = 0.97, d= 0.006), while we observed a trend of TD looking more to the face than DS (t = 1.7, p = 0.09, d = 0.37), and TD looking less to the background than DS (t = −1.68, p = 0.09, d = 0.36). This suggests that even when average latency in the visual orientation task is taken into account the same pattern of findings across groups holds in the audiovisual task.

3.3. Correlation between Audiovisual Task and Communicative Skills

Using the CSBS-DP we correlated concurrent communicative skills with infants' performance in the audiovisual task. For the current study we focused on skills that are relevant for infants' performance in audiovisual communication. Therefore, we analyzed data (raw scores) from the following scales: emotion and eye gaze, communication, and gesture. Nineteen TD infants (mean age 6 months, range 6–6 months), and six DS infants (mean age 6.8, age range 6–8 months) provided CSBS data. Note that for those infants that were younger than 6 months (i.e., the minimum assessment age for the CSBS questionnaire) at the moment of the AV task, the CSBS data were collected later (i.e., within the time span of 2–3 months for 2 of the infants). We observed that in the TD group there was a significant positive correlation between the proportion of looking to the eyes in the audiovisual task and the score on the gesture scale (r(18) = 0.47, p = 0.01), as well the communication score (r(18) = 0.36, p = 0.05). These results are depicted in Figure 4. In addition, we observed a marginal correlation between looking to the arm and the gesture score (r(18) = 0.34, p = 0.06). Other areas of interest did not provide a significant correlation with the CSBS scales (all ps > 0.1). Finally, for the DS group we observed no significant correlations (all ps > 0.2).

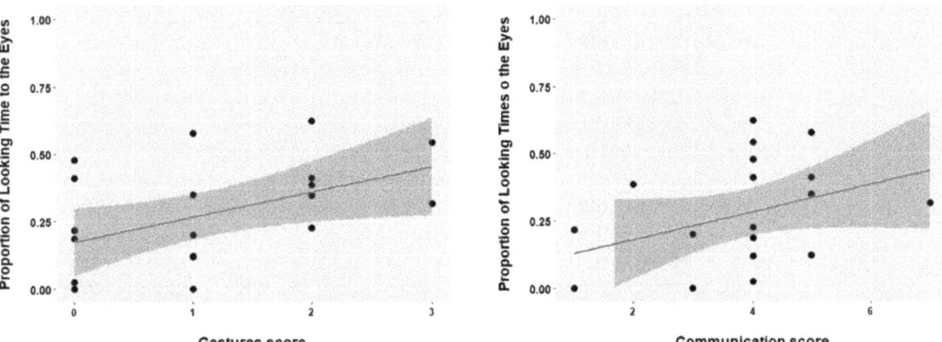

Figure 4. Scatter plot representing the relation between attention to the eyes in the audiovisual task, and gesture and communication score from the CSBS-DP for the TD group. Dots represent individual scores.

4. Discussion

The current paper focuses on early abilities that might be supporting language development in DS and TD 5–7-months old infants. In particular, we assessed infants' early visual attention, audiovisual speech processing, and communication skills. We will discuss each of the assessed measures and their implications for language development, particularly for DS infants. First, we observed that DS infants are slower in orienting their visual attention to stimuli in comparison to TD peers. This means that DS infants need more time to start attending to salient visual cues in their environment, here a flashing red light. This result is in line with previous studies on impaired visual attention in DS toddlers and children, especially in disengaging their visual attention [34–36]. However, our results differ from Landry and Bryson's study [35] where DS preschool children were similar in the visual orientation task to TD children matched in mental age with the DS group. There are at least two explanations for these between-studies differences. First, we tested a much younger population than in Landry and Bryson [35] and it is possible that by the preschool age DS children do improve their visual orientation attention. Second, there are important methodological differences between studies. Note that we compared DS and TD infants that were matched in their chronological, rather than in their mental age. Considering that we were interests in assessing DS infants between 5–7 months of age it would be difficult, if not impossible, to match groups in their mental age. Moreover, our task required infants to turn and orient their head to lateral/central position to flashing lights, whereas Landry and Bryson [35] used a set up with a central monitor and two lateral monitors placed in front of the child. It is therefore possible that our task was particularly challenging for the Down syndrome group. Nevertheless, our study is one of the first studies demonstrating that in first half of the first year of life, DS infants are impaired in orienting their visual attention to salient stimuli.

The second component we assessed was infants' attentional pattern during audiovisual speech processing. We presented an animated character that waived and talked at the infant. We were particularly interested in examining what visual speech/communicative cues infants attend to at this age early age. We observed that the two groups demonstrated certain similarities in their looking pattern: both groups do attend less to the waving arm at this age than the face and the background, suggesting that at this age the waiving gesture is not a particular salient communicative cue. Further, when attending to the face, both groups look more at the eyes than the mouth, in line with many recent studies done at similar ages (e.g., [41–43]). However, we observed a striking difference between the groups: TD infants attend more to the face than to the background, whereas DS infants attend similarly (~40 percent of their looking time to the screen) to the background and the face. This suggests that for DS infants the face is not a salient cue, at least not more salient than the background. Further implying that when observing a scene, DS infants attend equally to social/speech cues and to other cues that are not relevant for communication. This finding is in line with other studies that observed impairment in face and audiovisual speech processing in DS infants and toddlers [31–33]. Considering that we observed that DS infants are also slower to orient their visual attention, we can propose that such impairment affects their ability to orient to and fixate salient audiovisual speech cues (e.g., the face). Therefore, early interventions that are based on improving communication abilities have to take into account that DS attention to communicative cues is impaired. Moreover, we detected such impairment already in the first half of the first year of life, suggesting that interventions could target the ability to detect visual speech cues, even before interventions focusing on improving joint attention take place. We could also speculate that improving visual orientation might improve their ability to detect visual speech/communication cues and further research should address this possibility. It is important to note, however, that we have assessed audiovisual speech processing using an animated character, that certainly differs from a human face. In particular, the richness of the interplay between the acoustic signal, articulatory movements and facial expressions is reduced in animated characters, in comparison to a human face. Interestingly, even with the animated character,

DS and TD infants attend more to the eyes than the mouth, similar to previous studies assessing attention to a human face (e.g., [41–43]). Moreover, we also found that DS infants differ from TD infants in how they attend to communicative cues in an animated character talking face. It might be possible that for DS infants a human talking face would elicit greater attention, and future research should assess DS attention to a talking human face in relation to attractive background.

Finally, we observed important across-group differences regarding the relation between attending to audiovisual communicative cues and communicative development. First, we observed that attention to the eyes in the TD group relates to concurrent gesture and communication skills (and a trend for a positive correlation between looking to the arm and gesture skills was also found). Note that the items in the gesture and communication scales are mostly tackling non-verbal communication skills. For instance, assessing whether the child is pointing, waiving, asking for attention when a caregiver is not providing it, etc. So, we observed that in ~6-month-old typically developing infants, attending to relevant audiovisual cues, particularly to the eyes and the arm, supports early communication skills. This result is in line with previous studies on the importance of audiovisual cues for communication skills in typical development (e.g., [44]). However, in DS infants, we have not observed patterns supporting a relationship between performance in the audiovisual task and communication skills. We further inspected these results and observed that all DS infants have value 0 on the gesture score, meaning that gesturing in 5–7-month-old DS infants did not emerge yet. Considering that we have observed that DS infants attend much less to relevant audiovisual cues than TD infants, it is possible that less attention to audiovisual cues hinders their communicative skills. Alternatively, their poor communicative skills could drive away their attention from relevant audiovisual speech/communication cues. Either way, we observed that in the first 6 months of life, DS infants' link between audiovisual communicative attention and communication skills is not yet established. Further research is needed to address what is the relation between early attention to communicative cues and later communicative development.

A limitation of our study is certainly the small sample size, and future work should include larger samples. However, it should be noted that the population of DS infants is far less than that TD infants. According to the report from the National health institute [45] in the period from 2008–2017, in average, ~20 Down syndrome infants were born in Portugal per year. Thus, we assessed 35% of the population in a given year. An additional limitation of the current study is that our DS group was chronologically age matched with the TD group, and not mental age matched. Therefore, it is possible that the findings might change if groups were matched by mental age, in particular if an older group of DS infants was considered instead. However, that would leave our goal of investigating very early attention and audiovisual abilities in DS infants unaddressed, as well as of contributing to understand how DS communicative abilities develop from an early age. In future work, we plan to look at older DS infants to examine how these abilities develop. Nevertheless, based on the current findings we could speculate that undeveloped attentional orientation skills hinder DS infants' orientation to speech/communication cues. Therefore, future research should also explore intervention strategies that would focus on improving orientating attention, but specifically to audiovisual speech/communication cues, i.e., the face.

5. Conclusions

The current study assessed early attentional and audiovisual processing abilities in typically developing and in Down syndrome infants at 5–7 months of age. The study showed, for the first time, that at such an early age DS infants' attention and audiovisual speech processing is following a different developmental path than typically developing infants. We also observed that audiovisual attention supports concurrent communicative abilities in TD infants, but not in DS infants. In short, the current study demonstrated that early visual attention and audiovisual speech processing might be impaired in DS infants with consequences for their communication development, opening new avenues for early interventions in this clinical population. Furthermore, results from this study suggest that in face-to-face communication, DS infants might need more time to detect/attend to communicative cues, and caregivers might emphasize from early age face-to-face communication as a form of training attention to communicative cues.

Author Contributions: S.F. conceived and designed the study. J.P. and C.S. collected the data. J.P. and M.C. analyzed the data. J.P. and S.F. wrote the manuscript. All authors contributed to the article and approved the submitted version. All authors have read and agreed to the published version of the manuscript.

Funding: This research was funded by the Fundação para a Ciência e Tecnologia, Portugal (Grant PTDC/MHC-LIN/3901/2014, PI SF) in conjunction with the European Regional Development Fund from the EU, Portugal 2020 and Lisboa 2020 (Grant PTDC/LLT-LIN/29338/2017, PI SF). The APC was funded by Grant PTDC/LLT-LIN/29338/2017.

Institutional Review Board Statement: The study was conducted according to the guidelines of the Declaration of Helsinki, and approved by the Ethical Committee for Research (CEI) of the School of Arts and Humanities of the University of Lisbon (1_CEI2018).

Informed Consent Statement: Written informed consent to participate in this study was provided by the participants' legal guardian/next of kin.

Data Availability Statement: The raw data supporting the conclusions of this article will be made available by the authors, without undue reservation, on request to the corresponding author.

Acknowledgments: We would like to thank the research assistant Ricardo Sousa for help with data collection, and the audience of the NeuroDWELL workshop, Lisbon, 2019, for comments on a preliminary version of this study. We gratefully acknowledge the collaboration of the Center for Child Development Diferenças in recruitment of participants with Down syndrome, and of all the infants and their families.

Conflicts of Interest: The authors declare no conflict of interest.

References

1. Chapman, R.S. Language development in children and adolescents with Down syndrome. *Ment. Retard. Dev. Disabil. Res. Rev.* **1997**, *3*, 307–312. [CrossRef]
2. McDuffie, A.; Abbeduto, L. Language disorders in children with mental retardation of genetic origin: Down syndrome, Fragile X syndrome, and Williams syndrome. In *Handbook of Child Language Disorders*; Psychology Press: Hove, UK, 2009; pp. 44–66.
3. Abbeduto, L.; Warren, S.F.; Conners, F.A.; Sohail, A.; Ahmad, Z.; Ali, I. Language development in Down syndrome: From prelinguistic period to the acquisition of literacy. *Ment. Retard. Dev. Disabil. Res. Rev.* **2007**, *13*, 247–261. [CrossRef]
4. Finestack, L.H.; Sterling, A.M.; Abbeduto, L. Discriminating Down Syndrome and Fragile X Syndrome based on language ability. *J. Child Lang.* **2013**, *40*, 244–265. [CrossRef] [PubMed]
5. Kent, R.D.; Vorperian, H.K. Speech Impairment in Down Syndrome: A Review. *J. Speech Lang. Hear. Res.* **2013**, *56*, 178–210. [CrossRef]
6. Sommers, R.K.; Patterson, J.P.; Wildgen, P.L. Phonology of Down Syndrome Speakers, Ages 13–22. *Commun. Disord. Q.* **1988**, *12*, 65–91. [CrossRef]
7. Dodd, B.J. Comparison of bablibling pattehrns in normal and Down synfrome infants. *J. Intellect. Disabil. Res.* **1972**, *16*, 35–40. [CrossRef] [PubMed]
8. Steffens, M.L.; Oller, D.K.; Lynch, M.; Urbano, R.C. Vocal development in infants with Down syndrome and infants who are developing normally. *Am. J. Ment. Retard.* **1992**, *97*, 235–246. [PubMed]
9. Stoel-Gammon, C. Down syndrome phonology: Developmental patterns and intervention strategies. *Down Syndr. Res. Pract.* **2001**, *7*, 93–100. [CrossRef]
10. Stojanovik, V. Prosodic deficits in children with Down syndrome. *J. Neurolinguist.* **2011**, *24*, 145–155. [CrossRef]
11. Roberts, J.; Price, J.; Malkin, C. Language and communication development in Down syndrome. *Ment. Retard. Dev. Disabil. Res. Rev.* **2007**, *13*, 26–35. [CrossRef]
12. Mason-Apps, E.; Stojanovik, V.; Houston-Price, C.; Buckley, S. Longitudinal predictors of early language in infants with Down syndrome: A preliminary study. *Res. Dev. Disabil.* **2018**, *81*, 37–51. [CrossRef]
13. Hahn, L.J.; Loveall, S.J.; Savoy, M.T.; Neumann, A.M.; Ikuta, T. Joint attention in Down syndrome: A meta-analysis. *Res. Dev. Disabil.* **2018**, *78*, 89–102. [CrossRef]
14. Tomasello, M.; Farrar, M.J. Joint Attention and Early Language. *Child Dev.* **1986**, *57*, 1454. [CrossRef] [PubMed]
15. Cleveland, A.; Schug, M.; Striano, T. Joint attention and object learning in 5- and 7-month-old infants. *Infant Child Dev.* **2007**, *16*, 295–306. [CrossRef]

16. Baldwin, D.A.; Markman, E.M. Establishing Word-Object Relations: A First Step. *Child Dev.* **1989**, *60*, 381. [CrossRef] [PubMed]
17. Hirotani, M.; Stets, M.; Striano, T.; Friederici, A.D. Joint attention helps infants learn new words: Event-related potential evidence. *Neuroreport* **2009**, *20*, 600–605. [CrossRef] [PubMed]
18. Akhtar, N.; Gernsbacher, M.A. Joint Attention and Vocabulary Development: A Critical Look. *Lang. Linguist. Compass* **2007**, *1*, 195–207. [CrossRef]
19. Yu, C.; Suanda, S.H.; Smith, L.B. Infant sustained attention but not joint attention to objects at 9 months predicts vocabulary at 12 and 15 months. *Dev. Sci.* **2019**, *22*, e12735. [CrossRef]
20. Carpenter, M.; Nagell, K.; Tomasello, M.; Butterworth, G.; Moore, C. Social Cognition, Joint Attention, and Communicative Competence from 9 to 15 Months of Age. *Monogr. Soc. Res. Child Dev.* **1998**, *63*. [CrossRef]
21. Zampini, L.; Salvi, A.; D'Odorico, L. Joint attention behaviours and vocabulary development in children with Down syndrome. *J. Intellect. Disabil. Res.* **2015**, *59*, 891–901. [CrossRef]
22. Hahn, L.J. Joint Attention and Early Social Developmental Cascades in Neurogenetic Disorders. *Int. Rev. Res. Dev. Disabil.* **2016**, *51*, 123–152. [CrossRef]
23. Sumby, W.; Pollack, I. Visual Contribution to Speech Intelligibility in Noise. *J. Acoust. Soc. Am.* **1954**, *26*, 212. [CrossRef]
24. Król, M.E. Auditory noise increases the allocation of attention to the mouth, and the eyes pay the price: An eye-tracking study. *PLoS ONE* **2018**, *13*, e0194491. [CrossRef] [PubMed]
25. Weatherhead, D.; White, K.S. Read my lips: Visual speech influences word processing in infants. *Cognition* **2017**, *160*, 103–109. [CrossRef] [PubMed]
26. Teinonen, T.; Aslin, R.N.; Alku, P.; Csibra, G. Visual speech contributes to phonetic learning in 6-month-old infants. *Cognition* **2008**, *108*, 850–855. [CrossRef] [PubMed]
27. de la Cruz-Pavía, I.; Gervain, J.; Vatikiotis-Bateson, E.; Werker, J.F. Finding phrases: On the role of co-verbal facial information in learning word order in infancy. *PLoS ONE* **2019**, *14*, e0224786. [CrossRef]
28. Patterson, M.; Werker, J. Matching phonetic information in lips and voice is robust in 4.5-month-old infants. *Infant Behav. Dev.* **1999**, *22*, 237–247. [CrossRef]
29. Kuhl, P.; Meltzoff, A. The Intermodal Representation of Speech in Infants. *Infant Behav. Dev.* **1984**, *7*, 361–381. [CrossRef]
30. Pejovic, J.; Yee, E.; Molnar, M. Speaker matters: Natural inter-speaker variation affects 4-month-olds' perception of audio-visual speech. *First Lang.* **2019**, 1–15. [CrossRef]
31. Legerstee, M.; Bowman, T.G. The development of responses to people and a toy in infants with down syndrome. *Infant Behav. Dev.* **1989**, *12*, 465–477. [CrossRef]
32. Berger, J.; Cunningham, C.C. The development of eye contact between mothers and normal versus Down's syndrome infants. *Dev. Psychol.* **1981**, *17*, 678–689. [CrossRef]
33. D'Souza, D.; D'Souza, H.; Johnson, M.H.; Karmiloff-Smith, A. Audio-visual speech perception in infants and toddlers with Down syndrome, fragile X syndrome, and Williams syndrome. *Infant Behav. Dev.* **2016**, *44*, 249–262. [CrossRef]
34. D'Souza, D.; D'Souza, H.; Jones, E.J.H.; Karmiloff-Smith, A. Attentional abilities constrain language development: A cross-syndrome infant/toddler study. *Dev. Sci.* **2020**, *23*, 1–12. [CrossRef] [PubMed]
35. Landry, R.; Bryson, S.E. Impaired disengagement of attention in young children with austism. *J. Child Psychol. Psychiatry Allied Discip.* **2004**, *45*, 1115–1122. [CrossRef] [PubMed]
36. Breckenridge, K.; Braddick, O.; Anker, S.; Woodhouse, M.; Atkinson, J. Attention in Williams syndrome and Down's syndrome: Performance on the new early childhood attention battery. *Br. J. Dev. Psychol.* **2013**, *31*, 257–269. [CrossRef] [PubMed]
37. Frota, S.; Vicente, S.; Filipe, M.; Vigário, M. *CSBS DPTM Infant-Toddler Checklist from Communication and Symbolic Behavior Scales Developmental Profile-Portuguese Translation*; Prizant & Wetherby © 2002, and Portuguese Translation © 2014–2016; Translated by Permission Granted to the EBELa Project's Team; Paul, H., Ed.; Brookes Publishing Co., Inc.: Baltimore, MD, USA, 2018.
38. Meints, K.; Woodford, A. *Lincoln Infant Lab Package 1.0: A New Programme Package for IPL, Preferential Listening, Habituation and Eyetracking*; University of Lincoln: Lincoln, UK, 2008.
39. Frota, S.; Butler, J.; Uysal, E.; Severino, C.; Vigário, M. European Portuguese-Learning Infants Look Longer at Iambic Stress: New Data on Language Specificity in Early Stress Perception. *Front. Psychol.* **2020**, *11*, 1890. [CrossRef]
40. Kuznetsova, A.; Brockhoff, P.B.; Christensen, R.H.B. lmerTest Package: Tests in Linear Mixed Effects Models. *J. Stat. Softw.* **2017**, *82*, 1–26. [CrossRef]
41. Lewkowicz, D.J.; Hansen-Tift, A.M. Infants deploy selective attention to the mouth of a talking face when learning speech. *Proc. Natl. Acad. Sci. USA* **2012**, *109*, 1431–1436. [CrossRef] [PubMed]
42. Pejovic, J. *The Development of Audiovisual Vowel Processing in Monolingual and Bilingual Infants: A Cross-Sectional and Longitudinal Study*; University of the Basque Country: San Sebastian, Spain, 2019.
43. Cruz, M.; Butler, J.; Severino, C.; Filipe, M.; Frota, S. Eyes or mouth? Exploring eye gaze patterns and their relation with early stress perception in European Portuguese. *J. Port. Linguist.* **2020**, *19*, 1–13. [CrossRef]
44. Kushnerenko, E.; Tomalski, P.; Ballieux, H.; Ribeiro, H.; Potton, A.; Axelsson, E.L.; Murphy, E.; Moore, D.G. Brain responses to audiovisual speech mismatch in infants are associated with individual differences in looking behaviour. *Eur. J. Neurosci.* **2013**, *38*, 3363–3369. [CrossRef]
45. *Annual Report 2017: Boletim Epidemiológico Observações [Epidemiology Bulletin]*; Instituto Nacional de Saúde Doutor Ricardo Jorge [National Institute for Health Dr. Ricardo Jorge]: Lisbon, Portugal, 2017.

Article

The Association between Difficulties with Speech Fluency and Language Skills in a National Age Cohort of Children with Down Syndrome

Kari-Anne B. Næss [1,*], Egil Nygaard [2], Hilde Hofslundsengen [3] and J. Scott Yaruss [4]

[1] Department of Special Needs Education, University of Oslo, 0318 Oslo, Norway
[2] Department of Psychology, University of Oslo, 0373 Oslo, Norway; egilny@psykologi.uio.no
[3] Faculty of Teacher Education, Arts and Sports, Western Norway University of Applied Sciences, 6851 Sogndal, Norway; Hilde.Hofslundsengen@hvl.no
[4] Communicative Sciences and Disorders, Michigan State University, East Lansing, MI 48824, USA; jsy@msu.edu
* Correspondence: k.a.b.nass@uv.uio.no; Tel.: +47-922-40-741

Abstract: The present study (a) addressed difficulties in speech fluency in children with Down syndrome and typically developing children at a similar non-verbal level and (b) examined the association between difficulties with speech fluency and language skills in children with Down syndrome. Data from a cross-sectional parent survey that included questions about children's difficulties with speech fluency, as well as clinical tests from a national age cohort of 43 six-year-olds with Down syndrome and 57 young typically developing children, were collected. Fisher's exact test, Student's *t*-test, linear regression, and density ellipse scatter plots were used for analysis. There was a significantly higher occurrence of parent-reported difficulties with speech fluency in the children with Down syndrome. Higher language scores were significantly associated with a lower degree of difficulties; this association was strongest for vocabulary and phonological skills. Although difficulties with speech fluency were not reported for all children with Down syndrome, a substantially higher occurrence of such difficulties was reported compared to that for typically developing children. The significant association between difficulties with speech fluency and the level of language functioning suggests that speech fluency and language skills should be taken into consideration when planning treatment for children with Down syndrome.

Keywords: down syndrome; fluency; disfluency; co-occurrence; comorbidity; language

Citation: Næss, K.-A.B.; Nygaard, E.; Hofslundsengen, H.; Yaruss, J.S. The Association between Difficulties with Speech Fluency and Language Skills in a National Age Cohort of Children with Down Syndrome. *Brain Sci.* **2021**, *11*, 704. https://doi.org/10.3390/brainsci11060704

Academic Editor: Margaret B. Pulsifer

Received: 18 April 2021
Accepted: 22 May 2021
Published: 26 May 2021

Publisher's Note: MDPI stays neutral with regard to jurisdictional claims in published maps and institutional affiliations.

1. Introduction

A child's level of speech fluency can affect effective communication [1]. Disfluent speech is common in young children during periods when speech, language, and emotional functioning progress rapidly [2,3]. One group that is reported to exhibit difficulties with speech fluency across ages is individuals with Down syndrome [4–6]. Down syndrome is the most commonly known single biological cause of intellectual disability [7,8]; it affects more than 1 live birth per 1000 [9]. Considerable risk of communication and language disorder has been observed in previous research with this group of children [10,11]. Variables that may be associated with language disorder in this group of children include varying extents of hearing loss, including repeated "otitis media with effusion" [12–14]; oral and palate conditions [15,16], including differences in the craniofacial structures and shape of the palate and hypotonic oral musculature [17]; and reduced cognitive functioning [18], including impaired auditory short-term memory [19,20]. The language profiles of children with Down syndrome commonly show a relative gap in expressive versus receptive language skills, favouring the receptive domain (c.f. [21]). Consistent weaknesses, compared to typically developing children of similar non-verbal mental age level, are reported in the areas of expressive vocabulary, receptive and expressive grammar (syntax and morphology; [22–24]), and phonological awareness ([25,26]; see also a systematic review and meta-analysis by Næss et al. [11]). Speech production, including speech fluency, is also commonly affected [27–29]. Although there is an initial gap between expressive and receptive language domains and between vocabulary and other core language skills, all of these areas develop more slowly over time in children with Down syndrome than in younger typically developing peers with similar non-verbal mental age levels, and the gap between the groups increases over time [26,30,31].

Previous research shows that the level of language functioning (see, e.g., the review by Ntourou et al. [32]) and dissociations across language domains may relate to fluency difficulties in typically developing children [33,34]. The high co-occurrence of Down syndrome and disfluency, combined with the specific language profile in children with Down syndrome (which includes a low level of language functioning and a gap between expressive and receptive language domains), suggests that such a link may also exist for children with Down syndrome. This question has not been thoroughly investigated in previous research, however. Thus, in the present study, we aimed to investigate whether there is an association between difficulties with speech fluency and language functioning in children with Down syndrome.

1.1. Difficulties with Speech Fluency in Children with Down Syndrome

Disfluencies of different types may interrupt the smooth flow of speech [35]. Some of these disfluencies may reflect a communication disorder such as stuttering ("childhood onset fluency disorder" in the DSM–5; [1]). Examples of stuttering-like disfluency include repetitions of sounds or syllables, prolongations in sounds, or blocks [3,36]. Non-stuttering disfluencies, also called "other" disfluencies [27,37], are experienced by most speakers. These include interjections, repetitions of multisyllabic words or phrases, and revisions [3]. Research suggests that children with Down syndrome exhibit all types of disfluencies [6,27], and they show more frequent stuttering-like disfluencies than other types of disfluencies [27]. Research indicates a higher occurrence of stuttering within individuals with Down syndrome than in both typically developing individuals [38] and individuals with intellectual disability due to other causes [39]. Very few studies have directly compared difficulties with speech fluency between children with Down syndrome and typically developing children. Instead, studies have reported only the occurrence of fluency difficulties within a group of children with Down syndrome, or they have used results from other studies of typically developing children as reference values for comparison with their own values measured from children with Down syndrome.

Estimates of the prevalence and incidence of stuttering in the otherwise typically developing population vary between 5% [40] and 11% [41], while in children with Down syndrome, the prevalence of stuttering varies between 10% and 47% [5,42]. The large apparent variation in results across studies focusing on individuals with Down syndrome may be due, in part, to the differences in the consideration of the types of speech disfluencies (see review [27]), the criteria used for diagnosing stuttering (c.f. [43,44]), and the languages spoken (c.f. [45]). In addition, methodological issues, such as small samples of individuals with Down syndrome (e.g., $N = 28$ in [46]; $N = 26$ in [27]; $N = 1$ in [47]; $N = 5$ in [48]) or the wide age range of the participants, may have impacted the results. Notably, the practice of including both children and adults in the same study sample (e.g., age ranging from 3.8 years to 57.3 years; [6]; see also the review by Kent and Vorperian [42]) is problematic due to the phenotype of Down syndrome. For example, neuropathologies characteristic of Alzheimer's disease may already be pervasive in adults with Down syndrome by their 30s [49]. This may introduce a bias associated with the occurrence of difficulties with fluency, as language and communication are often reliably affected in this disease [50,51]. In particular, semantic verbal fluency has been found to be strongly associated with Alzheimer's disease in individuals with Down syndrome [52].

To our knowledge, very few previous studies have investigated the occurrence of difficulties with speech fluency in samples consisting *only* of children with Down syndrome. Eggers and van Eerdenbrugh [27] are, as far as we know, the only one (ages 3.03–12.06 years). Salihovic et al. [53], Schieve et al. [38], and Wilcox [48] also investigated speech fluency in children, but they all had a mixed sample with teenagers. Notably, mixing these age groups or even mixing preschool-age children and school-age children may introduce uncertainties into the data and make it difficult to discern the true occurrence of difficulties with speech fluency in the population of children with Down syndrome. For example, in typically developing children, a higher occurrence of children with difficulties with speech fluency is suggested in preschool-aged children than in school-aged children [41,54]. This means that the wide age range in previous studies and the common lack of a typically developing comparison group could have biased the occurrence estimates and evaluation of difficulties with speech fluency in children with Down syndrome.

1.2. The Purpose of the Present Study

Although the evidence regarding the occurrence of difficulties with speech fluency in children with Down syndrome has limitations, the existing research results are generally consistent across several studies: children with Down syndrome are more likely to exhibit disfluent speech than other children. Language disorders resulting from a lower level of language skills and dissociations between the receptive and expressive language domains are also more apparent in this group of children. Together, these patterns lead to the hypothesis that there is a potential association between language functioning and disfluency in children with Down syndrome. However, there are uncertainties about the role that language development may play in the speech fluency of individuals with Down syndrome. The potential relationship between language and disfluency has not been thoroughly investigated in a sample of children with Down syndrome. In the current study, therefore, we studied a national age cohort of children with Down syndrome (and a group of typically developing children with similar non-verbal mental age levels) to ask the following research questions:

(1) What is the occurrence of difficulties with speech fluency in a national age cohort of children with Down syndrome compared to that of a cohort of typically developing children at the same non-verbal mental age level?
(2) What is the association between difficulties with speech fluency and language skills in children with Down syndrome?
(3) Is there more dissociation in expressive and receptive language scores among children with Down syndrome who have difficulties with speech fluency compared to children with Down syndrome who have no difficulties with fluency?

Based on the uncertainties about the categorization of speech disfluency in individuals with Down syndrome in previous research, specifically whether the presence of disfluencies might reflect a fluency disorder such as stuttering [27,42], we focused on difficulties with speech fluency in general rather than the presumed diagnosis of a particular type of fluency disorder. In this way, the present data contribute to the overall understanding of the potential relationship between difficulties with speech fluency and language development in children with Down syndrome. Such information will contribute new knowledge related to assessment and treatment practices for children with cooccurring Down syndrome and difficulties with speech fluency.

2. Materials and Methods

The data included in this paper are original and obtained from a larger research project on language, reading, and communication skills in a national age cohort of 43 children with Down syndrome [26]. The Regional Committees for Medical and Health Research Ethics Sør-øst approved the study (reference ID: 19732), including the information letter and consent form, in advance.

2.1. Participants

The invitation letter and consent form were sent out by habilitation services across Norway to the registered parents of each child with Down syndrome who was scheduled to start school. Parents of 43 children with Down syndrome accepted the invitation and returned the consent form to the principal investigator. In their acceptance, they also confirmed that their child did not have any known comorbid diagnoses of autism spectrum disorder and that Norwegian was the child's first language. Of the 43 children with Down syndrome, two participants were excluded because of missing data regarding difficulties with speech fluency. Thus, the final sample consisted of 41 participants (21 boys and 20 girls) with a chronological age of $M = 75.79$ months ($SD = 3.57$ months) and a raw non-verbal mental ability score (block design) of $M = 12.32$ ($SD = 5.51$). In addition, parents of 57 typically developing children with similar non-verbal mental abilities accepted the invitation to serve as controls. These children were recruited from eight kindergartens in a Norwegian municipality; they were required to have Norwegian as their first language and no history of special educational needs. Of the 57 typically developing children, 3 participants were excluded due to missing data on the dependent variable. The final sample of typically developing children consisted of 54 participants (26 boys and 28 girls; chronological age: $M = 36.50$ months, $SD = 4.15$ months; non-verbal mental ability raw score (block design): $M = 12.57$, $SD = 4.48$).

2.2. Data Collection

Two sources of data collection were used: a parental questionnaire administered online and clinical tests. For the parental questionnaire, an email was sent to one parent of each participating child. Two reminders were sent out if no answers were received within the deadline. The answers were automatically coded in SPSS from the digital questionnaire. For the clinical test data, children were assessed individually and in person in three sessions. All test answers were registered manually in the standardized test protocol, and expressive answers were audio recorded for later verification.

2.3. Measures

All measures included in this sub-study are presented below. For all tests, standardized procedures for implementation and scoring were followed. In the scoring of expressive tests, the children were not penalized for systematic articulation mistakes. Internal consistency as a function of the number of test items and the average intercorrelation among the items from the full sample of 43 children participating in the main project are reported in brackets in the individual test descriptions.

2.3.1. Difficulties with Speech Fluency/Stuttering

The dependent variable was assessed via the parent questionnaire language and reading development in children with Down syndrome (for a full English version of the questionnaire, see [55]). The parent was asked to rate their child's "degree of difficulties with speech fluency/stuttering," with a four-category answer option: from no difficulties with disfluency/stuttering (1) to a high degree of difficulties with speech fluency/stuttering (4). Difficulty with speech fluency/stuttering (hereafter called difficulties with speech fluency) was mainly analysed as a continuous variable, though it was dichotomized to investigate the last research question with "none" interpreted as indicating no difficulties with speech fluency and little, moderate, and high interpreted as indicating a difficulty with speech fluency.

2.3.2. Non-Verbal Mental Ability

We used the Block Design subtest of the Wechsler Preschool and Primary Scale of Intelligence (WPPSI-III; [56]) as a background measure. In this non-verbal test, the child was asked to copy a building block pattern (shown with blocks or as a picture). The maximum score was 40, and the internal consistency of the block design was high across all 20 items (Cronbach's $\alpha = 0.81$).

2.3.3. Vocabulary

Two tests of vocabulary were used: the Norwegian versions of the British Picture Vocabulary Scale (BPVS-II; [57,58]) and Picture Naming (WIPPSI–III; [56]). The BPVS-II is a receptive vocabulary test in which the examiner says a target word and the child is asked to point out the picture corresponding to the word among four pictures. The target words name animals, emotions, and professions, with increasing difficulties. The maximum score was 144, and the internal consistency of the BPVS was high across all 144 items (Cronbach's $\alpha = 0.93$). In the expressive vocabulary test, Picture Naming, the child was shown a picture of, e.g., a ball, a pencil, and an ambulance, and asked to name the item. The maximum score was 38, and the internal consistency of Picture Naming was high across the 38 items (Cronbach's $\alpha = 0.90$).

2.3.4. Grammar

Two tests were used to assess grammar: the Norwegian versions of the Test for Reception of Grammar (TROG-R; [59,60]) and Grammatic Closure (Illinois Test of Psycholinguistic Abilities (ITPA; [61,62]). In the receptive test TROG-R, the examiner says a sentence, and the child is asked to point out the picture that corresponds best among four pictures. The maximum score was 80, and the internal consistency of the TROG was high across all 80 items (Cronbach's $\alpha = 0.85$).

The Grammatic Closure test is an expressive subtest from the ITPA, where the child must answer grammatically correct nouns, verbs, adjectives, prepositions, and possessive pronouns. For example, if the child looked at a picture, and the examiner read a corresponding 'model' sentence: "Here is one bed. Here are two . . . ?", the child's task would be to finish the new sentence based on the 'model' sentence. The maximum score was 33, and the internal consistency of the Grammatic Closure was high across all 33 items (Cronbach's $\alpha = 0.72$).

2.3.5. Phonology

To assess phonological awareness, we used four receptive measures adapted from Carroll et al. [63]: initial syllable matching, final syllable matching, rhyme matching, and initial phoneme matching. In each of the tasks, a puppet was used to make the assessment more child friendly. For example, when assessing the initial syllable, the child was told that the puppet likes to collect words that start with the same syllable. The puppet showed a picture card to the child and asked the child to point at the picture that began with the same syllable among two more picture cards on the table. The task was presented in the same way for the other three phonological awareness measures, with a different puppet for each measure. The maximum scores of both syllable measures are 8 each, while the maximum scores of the remaining two measures are 16 each. Cronbach's α for initial syllable = 0.57, final syllable = 0.77, rhyme = 0.85, and phoneme = 0.83.

In addition to phonological awareness, expressive phonology was measured using a Norwegian version of the Children's Test of Non-Word Repetition [64,65]. In this test, the child first heard a non-word of between two and five syllables in length and was asked to repeat the word. The maximum score was 28, and the internal consistency of the Children's Test of Non-Word Repetition across all 28 items was high (Cronbach's $\alpha = 0.83$).

2.3.6. Speed of Processing

Speed of processing was measured by rapid automatized naming (RAN) and Child Language and Learning's speed of processing tests [66]. Two tasks with black and white drawings of objects were used to assess expressive processing speed. The objects represented high-frequency words usually acquired at a very early age, such as RAN1: a sun, a boat, a mouse, a door, and a bus; and RAN2: a light, a ball, a boy, a house, and a car. The five pictures were shown randomly in four rows with five items in each row. The child was asked to name each picture. The time it took to complete the task was recorded, and mean summary scores were calculated for the total amount of time used on the two tasks. The intraclass correlation between RAN1 and RAN2 was moderate (ICC = 0.57; when using a two-way mixed effects model with absolute agreement based on an average of the two measures [67].

Two tasks were used to assess the receptive speed of processing: both involved focusing on objects; the words used were high frequency and usually acquired at an early age [66]. In the first speed of processing task (SPEED 1), the child was given a sheet of paper showing black and white drawings of a sun, a boat, a mouse, a door, and a bus. The five pictures were shown randomly in six rows with seven items in each row. There were four sheets in total. For each sheet, the child was shown a mouse and asked to collect all the mice on the sheet. Then, the child was given a marker and asked to set a dot on all the mice on the sheet. Finally, the child was asked to do the task as quickly as possible for one minute. The number of tasks that the child completed correctly within the time frame was summarized. In the second speed of processing task (SPEED 2), the child carried out the same task as in SPEED 1, but the pictures were of a light, a ball, a boy, a house, and a car, and the child was asked to collect cars. The scoring scheme for SPEED 2 was the same as that for SPEED 1. The mean summary scores were calculated for the total amount of time taken to complete the two speed tasks. The intraclass correlation between SPEED 1 and SPEED 2 was good (ICC = 0.87; when using a two-way mixed effects model with absolute agreement based on an average of the two measures).

2.4. Analysis

In total, six values out of 820 possible scores (0.7%) for the predictors used in the regression analyses were missing among the children with Down syndrome on the SPEED tasks because some children did not want to do those tasks. The results from Little's test (chi-square (35) = 31.24, p = 0.65) indicated that these missing data were random, so the missing data were replaced by multiple imputation (50 datasets). All analyses and results, except for descriptive statistics (Table 1), were based on the data set that included these imputed data.

Table 1. Descriptive statistics.

	Children with No Disfluency (n = 12)			Children with Disfluency (n = 29)			Test of Difference (t-Test)
	Mean	*SD*	**Range**	**Mean**	*SD*	**Range**	*p*-**Value**
Non-verbal mental functioning	13.00	5.19	4–22	12.03	5.70	0–22	0.616
BPVS	28.33	13.06	2–50	20.69	10.33	1–37	0.053
Picture naming	10.17	5.44	1–16	7.48	5.65	0–20	0.170
TROG	9.25	6.61	1–27	8.62	5.07	0–19	0.743
Grammatic closure	2.08	2.54	0–6	1.10	1.74	0–6	0.161
Phonological awareness	24.17	8.54	5–41	17.24	10.79	0–31	0.055
Non-word repetition	3.17	2.48	0–7	2.55	3.50	0–10	0.584
Speed	9.17	5.32	0–17.50	7.55 [1]	5.73	0–22.50	0.413
RAN	49.26	43.44	0–179.5	48.65	27.73	0–103.50	0.957

Note [1] n = 27. Non-verbal mental functioning was assessed with the Block Design subtest. Language functioning was assessed with British Picture Vocabulary (BPVS) and Picture Naming for vocabulary, the Test for Reception of Grammar (TROG) and Grammatic Closure subtest from ITPA for grammar, sum of four Phonological awareness tests and the Children's Test of Non-Word Repetition for phonology skills, and the mean of two Child Language and Learning's speed of processing tests (Speed) and the mean of two Rapid Automized Naming tasks (RAN) for processing speed.

For the first research question, Fisher's exact test, Student's t-test, and linear regression analyses were used to test differences between children with Down syndrome and typically developing children. For the second research question, we combined receptive and expressive functioning within four domains: vocabulary, grammar, phonology, and processing speed. The associations between these four functional linguistic domains and the degree of difficulties with fluency were analysed with three levels of linear regression models: a bivariate model, a model controlling for non-verbal mental functioning and a full model including all four functional linguistic domains and non-verbal mental abilities as predictors. All variables were standardized (Z-values) before being combined, and all variables were again standardized before being entered into the regression models. Thus, the presented regression coefficients can be interpreted as standardized regression coefficients. For the third research question, we created disparity variables in which expressive functioning scores were subtracted from receptive functioning scores within each of the four functional domains (vocabulary, grammar, phonology, and processing speed). A total disparity variable was also calculated across all four domains. Again, all variables were standardized before being deducted, and the combined variables were standardized before being entered into linear regression analyses. The disparity variables were analysed as predictors for the degree of difficulties with fluency in bivariate analyses, controlled for non-verbal mental functioning, and in a full linear regression model with all functional linguistic domains and non-verbal mental functioning entered as independent variables. In addition, we investigated whether the confidence intervals for the regression coefficients for receptive and expressive functioning overlapped when they were entered separately into the model instead of the disparity variable. We used the Lmatrix function in general linear models to investigate whether the regression coefficients of receptive and expressive functioning were significantly different. We also analysed whether there is more dissociation in expressive and receptive language scores among children with Down syndrome who have difficulties with fluency compared to children with Down syndrome who have no difficulties with fluency in a similar manner to that done by Anderson et al. [34]. Specifically, we used density ellipse scatter plots to identify participants outside the 95% ellipse who also had a dissociation of more than 1 standard deviation between receptive and expressive scores.

The distribution of data was evaluated by analysing the residuals of the final regression models through histograms, scatterplots, and P-P plots. Multicollinearity was investigated through a correlation matrix and by the variance inflation factor (VIF). All analyses were performed using IBM SPSS Statistics version 27, with the exception of density ellipse plots, which were made with the package ggplot2 using R version 4.0.3. A significance level of 5% was chosen for all analyses. No a priori correction for multiple comparisons was made due to this being an exploratory observational study [68].

3. Results

3.1. Descriptive Statistics

The sex distribution was quite similar between samples, with 20 (49%) girls in the group with children with Down syndrome and 28 (52%) in the group of typically developing children (chi-square 0.088, $p_{\text{exact}} = 0.84$). There were no significant differences in children's non-verbal mental functioning ($M = 12.3$, $SD = 5.5$ and $M = 12.6$, $SD = 4.5$ for children with Down syndrome and typically developing children, respectively, $t(93) = 0.25$, $p = 0.80$). Descriptive language data for the children with Down syndrome are presented in Table 1.

3.2. Research Question 1: Occurrence of Difficulties with Fluency in Children with Down Syndrome and Typically Developing Children

The distribution of the parent-reported difficulty with speech fluency is presented in Table 2. If dichotomizing the symptoms, 29 (71%) of the children with Down syndrome were judged to have difficulties with speech fluency, compared to 8 (15%) of the typically developing children. The difference in the difficulties with fluency between children with Down syndrome and typically developing children was highly significant, independent of whether levels of symptoms were dichotomized (chi-square = 30.65, $p < 0.001$) or were used as continuous variables, both before ($\beta = 0.62$, $p < 0.001$) and after ($\beta = 0.61$, $p < 0.001$) controlling for non-verbal mental functioning.

Table 2. Degree of difficulties with fluency among children with Down syndrome and typically developing children at the same non-verbal mental age level.

Degree of Difficulties	Children with Down Syndrome (*n* = 41) Mean (*SD*)		Typically Developing Children (*n* = 54) Mean (*SD*)	
	n	%	*n*	%
None	12	29	46	85
Small	10	24	8	15
Moderate	13	32		
Severe	6	15		

3.3. Research Question 2: The Association between Difficulties with Fluency and Language Skills in Children with Down Syndrome

To investigate the association between language skills and difficulties with speech fluency, we created four variables representing the four functional linguistic domains of vocabulary, grammar, phonology, and processing speed. Each variable reflected the mean of standardized (Z) values of the receptive and expressive tests for each domain. Table 3 presents the results from bivariate linear regression analyses for each of these skills when controlling for non-verbal mental functioning and a full model with both non-verbal mental functioning and all four functional linguistic domains as predictors. Vocabulary skills were significantly related to difficulties with speech fluency in all models, with moderate [69] effect sizes (β between 0.52 and 0.61). Grammar, phonology, and processing speed had small to moderate effect sizes in bivariate analyses (β between 0.30 and 0.40) and when controlled for non-verbal mental functioning (β between 0.26 and 0.38). However, of the three, only phonology skills were significant when controlling for non-verbal mental functioning. The effect sizes for grammar, phonology, and processing speed were negligible when all four domains were included in the model. There were no indications of any violations of assumptions for linear regression analyses for the full model, and the highest VIF was 3.1, indicating that there was not a high degree of collinearity. Nevertheless, correlations between the four functional domains (r from 0.36 to 0.72) may have influenced the results in the full model (see correlation matrix in Table S1). As indicated above, no a priori correction for multiple comparisons was made. Nevertheless, the effects of vocabulary and phonology found in Table 3 are still significant after controlling for four comparisons [70].

Table 3. Regression analyses of association with difficulties with fluency among children with Down syndrome ($N = 41$).

	Bivariate Model				Controlled for Non-Verbal Mental Functioning				Full Model			
	B	95% CI of *B*	*SE*	*p*	*B*	95% CI of *B*	*SE*	*p*	*B*	95% CI of *B*	*SE*	*p*
Vocabulary	−0.52	−0.79, −0.26	0.14	<0.001	−0.60	−0.93, −0.28	0.17	<0.001	−0.61	−1.10, −0.11	0.25	0.016
Grammar	−0.32	−0.61, −0.02	0.15	0.038	−0.31	−0.69, 0.07	0.19	0.109	0.06	−0.38, 0.50	0.22	0.777
Phonology	−0.40	−0.68, −0.11	0.15	0.007	−0.38	−0.70, −0.06	0.16	0.020	−0.07	−0.51, 0.38	0.23	0.776
Processing speed	−0.30	−0.60, 0.00	0.15	0.053	−0.26	−0.59, 0.07	0.17	0.127	0.03	−0.35, 0.41	0.19	0.879

Note. The results from regression analyses were all standardized (Z-values) before being entered into the models. Thus, *B* can be interpreted as a standardized regression coefficient. The results for intercepts and non-verbal mental functioning are not shown, as these are not the subject of the present article. The full model includes non-verbal mental functioning and four variables combined from both receptive and expressive features: vocabulary, grammar, phonology abilities, and processing speed. Non-verbal mental functioning was assessed with the Block Design subtest. Receptive and expressive functioning were assessed with the British Picture Vocabulary and Picture Naming for vocabulary, the Test for Reception of Grammar and Grammatic Closure subtest from ITPA for grammar, the mean of four Phonological awareness tests and the Children's Test of Non-Word Repetition for phonological skills, and the Child Language and Learning's speed of processing tests and the Rapid Automized Naming task for processing speed. Data are based on 50 multiple imputed datasets for 0.7% missing data.

3.4. Research Question 3: The Dissociation in Expressive and Receptive Language Scores between the Groups

To investigate the dissociation in expressive versus receptive functioning among children with Down syndrome, we created new variables presenting the discrepancy between receptive and expressive functioning. To ensure comparability across measures, all variables were standardized (Z-values) before expressive functioning was deducted from receptive abilities. No violation of assumptions for regression analyses or collinearity were found. Table 4 presents the association between these differences and the level of difficulties with fluency. Neither of the investigated domains of vocabulary, grammar, phonology, or processing speed nor the total receptive versus expressive difference were related to the level of difficulties with fluency.

Table 4. Association between difficulties with fluency and the gap between receptive and expressive skills in various language areas ($N = 41$).

Language Area	Bivariate Model				Controlled for Non-Verbal Mental Functioning				Full Model			
	B	95% CI of *B*	*SE*	*p*	*B*	95% CI of *B*	*SE*	*p*	*B*	95% CI of *B*	*SE*	*p*
Vocabulary	−0.03	−0.34, 0.28	0.18	0.850	−0.02	−0.33, 0.29	0.16	0.899	0.03	−0.33, 0.40	0.18	0.852
Grammar	0.07	−0.25, 0.38	0.16	0.678	0.10	−0.21, 0.41	0.16	0.534	0.19	−0.16, 0.55	0.18	0.280
Phonology	−0.21	−0.51, 0.10	0.16	0.191	−0.21	−0.51, 0.10	0.16	0.177	−0.30	−0.66, 0.06	0.19	0.104
Processing speed	−0.06	−0.37, 0.26	0.16	0.731	0.02	−0.32, 0.35	0.17	0.923	−0.05	−0.40, 0.29	0.18	0.762
Total	−0.04	−0.35, 0.28	0.16	0.820	0.00	−0.32, 0.32	0.16	0.993				

Note. Linear regression analyses of the association between difficulties with fluency and the differentiation between receptive and expressive functioning (receptive minus expressive). All variables were standardized before being combined, and all variables were again standardized before being entered into the regression models. The full model includes non-verbal mental functioning and four variables with the differentiation between receptive and expressive features: vocabulary, grammar, phonology, and processing speed. Non-verbal mental functioning was assessed with the Block Design subtest. Receptive and expressive functioning were assessed with British Picture Vocabulary and Picture Naming for vocabulary, the Test for Reception of Grammar and the Grammatic Closure subtest from the ITPA for grammar, the mean of four Phonological awareness tests and the Children's Test of Non-Word Repetition for phonological skills, and the Child Language and Learning's speed of processing tests and the Rapid Automized Naming task for processing speed. The total is the combination of the previous four domains. Data are based on 50 multiple imputed datasets for 0.7% missing data.

In addition, we investigated whether the regression coefficients for receptive versus expressive functioning overlapped when both were entered into linear regression analyses. In this analysis, non-verbal mental functioning was controlled for, and the level of difficulties with fluency was the dependent variable. For all five comparisons, the confidence intervals for receptive and expressive functioning highly overlapped. Thus, none of the five contrasts were significant when comparing the receptive and expressive regression coefficients after controlling for non-verbal mental functioning using the transformation coefficients matrix (MMATRIX) function in general linear models (GLMs).

We further analysed the material in a similar manner as previously done by Anderson et al. [34]. Table 5 gives an overview of cases that met both requirements for dissociation. There was a general tendency for more dissociation among children with low levels of difficulties with fluency than among children with moderate or high levels of difficulties, but this was only significant for grammar ($p = 0.05$).

Table 5. Cases with dissociative scores outside the density ellipse ($N = 41$).

	Children with No Difficulties with Fluency ($n = 12$)		Children with Difficulties with Fluency ($n = 29$)		Test of Difference Chi-Square
	n	%	n	%	p-Value
Vocabulary	0	0%	3	10%	0.543
Grammar	4	33%	2	7%	0.050
Phonology	1	8%	2	7%	1.00
Processing speed	1	8%	0	0%	0.293
Total	1	8%	2	7%	1.00

Note. Dissociation was defined as a case in which two requirements were met: (1) being outside a 95% density ellipse of the scatterplot between receptive and expressive functioning scores and (2) having a difference of more than 1 standard deviation in the two scores. Receptive and expressive functioning were assessed with British Picture Vocabulary and Picture Naming for vocabulary, the Test for Reception of Grammar and the Grammatic Closure subtest from the ITPA for grammar, the mean of four Phonological awareness tests and the Children's Test of Non-Word Repetition for phonological skills, and the Child Language and Learning's speed of processing tests and the Rapid Automized Naming task for processing speed. The total is the combination of the previous four domains. Data are based on 50 multiple imputed datasets for 0.7% missing data. p-values are based on Fisher's exact chi-square test.

4. Discussion

The results showed a significantly higher occurrence of difficulties with speech fluency in children with Down syndrome than in typically developing children with similar non-verbal mental age levels (corresponding to a chronological age of ca. 3 years). In addition, a large percentage of children with Down syndrome were rated as having serious difficulties with speech fluency. This stands in contrast to the finding that none of the typically developing children showed a serious degree of difficulties with speech fluency. The associations between language measures and the degree of difficulties with fluency in the children with Down syndrome were significant for all language domains included in the bivariate analysis; higher language skills were associated with a lower degree of difficulties with fluency. After taking into account non-verbal mental abilities, vocabulary and phonological skills were still significantly associated with the degree of difficulties with speech fluency. However, the dissociation hypothesis, that is, that there is a relationship between more fluency difficulties and larger gap between expressive and receptive language domains, was not supported by the data.

4.1. High Occurrence of Children with Difficulties with Speech Fluency

The high occurrence of difficulties with speech fluency in children with Down syndrome compared to typically developing children at the same non-verbal mental age level was expected based on inferences drawn from existing research. No previous studies have exactly investigated the occurrence of difficulties with speech fluency in children with Down syndrome compared to typically developing children at the same non-verbal mental age level. However, our results pattern align with the results from a survey study that used a group comparison design to investigate the occurrence of fluency disorders. Schieve et al. [38] included a sample of 27 individuals with Down syndrome and 1393 typically developing individuals and found occurrences of 15.6% and 1.5%, respectively. The results also align with results from an audio sample study comparing separate estimates of the occurrence in individuals with Down syndrome to estimates in previous research on the occurrence in typically developing individuals [27]. They found an occurrence of 31% in children with Down syndrome, which stands in stark contrast to the commonly cited values of a 1% prevalence of stuttering in typically developing individuals [35] and a lifespan incidence of more than 5% [71].

In addition to aspects of speech and language skills, various other developmental aspects of physical abilities and psychological state have been suggested to explain unique variance in the development of stuttering in otherwise typically developing children [2,72]. Nevertheless, such multifactorial models were not developed to explain difficulties with speech fluency in general, and they were not designed for unique populations such as children with Down syndrome. It is apparent from prior research that children with Down syndrome have challenges with all of these aspects of development, including language [26,73], speech motor skills [16,30], and emotionality (e.g., [74,75]). The complex developmental profile found in children with Down syndrome, including a range of different challenges, may also make these children vulnerable to developing difficulties with speech fluency.

4.2. More Serious Difficulties with Speech Fluency

The degree of parent-reported difficulties with speech fluency in our sample of children with Down syndrome varied from no difficulty to severe difficulty. For typically developing children, parents reported only no difficulty or a small degree of difficulty. These results imply more variation in the difficulties across children with Down syndrome than in the (younger) typically developing children. These results are in line with the results from Eggers and Van Eerdenbrugh [27], who also showed a large variation in the percentage in both stuttering-like disfluencies and other disfluencies across their sample of children with Down syndrome. However, in the current study, the differences in chronological age between the two participant groups may have influenced the results. This is because the parents may have rated their child's difficulties with fluency with peers at similar chronological age in mind. Due to the age effect of difficulties with fluency [71,76], the parents of the (older) children with Down syndrome may have an expectation of fewer difficulties with fluency in their children than the parents of the (younger) typically developing children whose age-matched peers may also have more disfluencies.

In the current study, some children with Down syndrome were rated to have *no* difficulties with speech fluency. In contrast, Eggers and Van Eerdenbrugh [27] reported that all of the children in their sample showed some disfluency. This difference in results may reflect that the current study focused on *difficulties* with speech fluency, while the study by Eggers and Van Eerdenbrugh [27] focused on the presence of a range of different types of speech disfluencies. This would allow a child to exhibit disfluencies without being judged by the parent to experience *difficulties* with fluency.

4.3. An Association between Difficulties with Speech Fluency and the Level of Language Skills

The results of the current study showed that better language skills are associated with a lower degree of difficulties with speech fluency. To the best of our knowledge, no other studies have investigated the association between language level and difficulties with speech fluency in *children* with Down syndrome. However, in typically developing children, associations have been reported between language skills and disfluency [77] and between language skills and stuttering [33,78–80]. Luckman et al. [79] found that children who stuttered scored almost one standard deviation below children who did not stutter on expressive vocabulary. In a range of studies, increasing the length and complexity of utterances has been found to be associated with increased stuttering in children [34,81–93]. Children who stutter are also shown to have increased difficulties with fluency on both monosyllabic function words [94] and unfamiliar words (non-words/novel phonological sequences) [95].

Children with Down syndrome usually have a broad language disorder affecting both sentence-level and word-level production. On average, they reach the milestone of sentence production at approximately 3.5 to 5 years of age [96,97], but 30% of children with Down syndrome still do not speak in complete sentences by the age of 6 [29]. For children who do speak in sentences, limitations in syntax and complex sentence structure are still reported [29,98], and they also have a low mean length of utterance [99]. In general, children with Down syndrome have limited expressive vocabulary [30] and show initial weaknesses in function words such as prepositions, conjunctions, and pronouns [100], as well as on unfamiliar words [24,26,101]. Children with Down syndrome may therefore be specifically vulnerable to difficulties with fluency due to aspects related to their expressive language skills, even though they may also be at a stage in their language development when they are still producing relatively simple sentences.

The association between difficulties with speech fluency and language in typically developing children who stutter has been the focus of a longstanding debate (e.g., [32,102, 103]). The fact that language is a common active ingredient in existing treatment programs for stuttering [104] also suggests an association between difficulties with speech fluency.

4.4. More Difficulties with Speech Fluency Not Related to Higher Level of Dissociation in Expressive and Receptive Language Skills

The children with Down syndrome in this study had, on average, both a low language level and a dissociation between expressive and receptive domains. Nevertheless, the results do not indicate an association between a higher degree of difficulties with speech fluency and a larger gap between expressive and receptive language skills, with the only marginally significant finding actually going in the opposite direction. Contradictory to our findings, studies with typically developing children suggest that gaps in performance within or between linguistic subcomponents, such as between receptive and expressive vocabulary, are associated with stuttering [33,34,83,105]. Anderson et al. [34] concluded that their sample of 45 children who stutter (age 3–5.11 years) was three times more likely to have dissociations across speech-language domains than their sample of 45 children who do not stutter (age 3–5.11 years). Coulter et al. [83] replicated the paper from Anderson et al. [34], and their results showed that children who stutter were five times more likely to have dissociations than children who do not stutter. They suggested that the dissociations could be markers of speech and language production systems that are not congruent with each other [83]. However, we did not find a relationship between higher level of dissociation and more difficulties with speech fluency in children with Down syndrome. This may be due to a minimal impact of the gap between receptive and expressive skills on difficulties with speech fluency or to the coarse assessment of difficulties with fluency and the parent-reported nature of the variable.

4.5. Limitations

A number of limitations of this study should be mentioned. First, the sample size of the study reduced the number of associated variables that could be included in the analysis. Although this is a relatively large study on difficulties with fluency in children with Down syndrome, the number of participants was still low for statistical analysis. To reduce the possible bias from adding too many covariates into the analysis, we summed the scores of two or more variables, but this may have the unintended effect of diluting the relevant contributions of individual variables.

To keep the sample as large as possible and to investigate an unselected sample of children with Down syndrome, no selection criteria were imposed to facilitate convenience in the recruitment process in the current study. Eggers and Van Eerdenbrugh [27] had an original sample of 50 participants, but the number of participants reported in their paper was 26. Their selections may have been based on the number of utterances/syllables available for each child. Obtaining a sufficiently large speech sample may be difficult in this clinical group, particularly when seeking to obtain enough syllables of speech to decide whether children stutter or not. Notably, professional coding of the disfluencies of children with Down syndrome may be challenging due to large variations in the speech produced and their phenotypic characteristics, including pauses and varying speech rates [106–109]; difficulties with prosody, including differences in lexical stress, producing questioning intonation, and the use of imitating intonation [73]; and articulation difficulties [28,110]. The present results therefore complement the results from Eggers and Van Eerdenbrugh [27] by adding information about *parental* judgements, which consider context and experiences. Tumanova et al. [3] highlighted that parents' report of difficulties with fluency in typically developing children is usually valid. On the other hand, it may be hard for parents of children with Down syndrome to evaluate their child's difficulties with fluency independent of their child's other complex speech, language, and communication disorders. Consequently, parents may have responded about language skills more generally than would an expert in speech and language therapy. The significant relationship between difficulties with fluency and language skills may therefore have been influenced by parents not clearly separating these two issues. Future research can supplement the knowledge base further by combining both parental and clinician judgements in evaluating the difficulties with fluency in children with Down syndrome.

The use of parental reports of difficulties with fluency status gives information about the difficulties with speech fluency across settings and partners but has limitations due to a lack of information about the severity and the types of disfluency—and whether the difficulties reflect an actual fluency disorder. In addition, it has been suggested that the difficulties with speech fluency in children with Down syndrome may represent a specific disfluency profile that does not fully overlap with the distribution of disfluencies in typically developing children who stutter [27]. This study adds knowledge related to difficulties with the fluency of a national age cohort and its association with language skills. It also confirms that parents' judgements are able to identify variations in the degree of difficulties with fluency within a sample of children with Down syndrome. However, checking the data against clinical judgements will be of importance, and future studies should ensure the inclusion of a set of measures in addition to the parent's judgements. Examples of data to be included in such inquiries include a clinical evaluation of typology, frequency, and severity of disfluencies across different speaking situations. There may also be some effects from fluency and/or language treatment that should be considered when the occurrence of difficulties with fluency are investigated in future research.

5. Conclusions

The results of parental data from this national age cohort of children with Down syndrome within a narrow age range indicate a significantly higher occurrence of difficulties with speech fluency compared to typically developing children of the same non-verbal mental age level. A significant association between difficulties with speech fluency and the level of language skills was discovered and should be taken into consideration when planning treatment for children with Down syndrome.

To date, limited research results on interventions and treatment of difficulties with fluency or fluency disorders for children with Down syndrome exist, and no effect study (e.g., a randomized controlled trial) is known to the authors. Until we know more about what constitutes effective treatment for this group of children, the large co-occurrence between difficulties with fluency and low language skills in children with Down syndrome supports a need for speech and language therapy that aims to simultaneously improve the child's language development and speech fluency. Speech-language pathologists also have a responsibility for treating the complex communication disorder of this group of children. To tailor the treatment to research-based knowledge, future effect studies should be designed, especially for children with Down syndrome; these studies should control for language level to investigate the potential effects of fluency treatment on language development.

Supplementary Materials: The following are available online at https://www.mdpi.com/article/10.3390/brainsci11060704/s1, Table S1. Correlation matrix (N = 41).

Author Contributions: Conceptualization, K.-A.B.N., E.N., H.H., J.S.Y.; methodology, K.-A.B.N., E.N., H.H.; formal analysis, E.N., K.-A.B.N.; investigation, K.-A.B.N.; resources, K.-A.B.N.; data curation, K.-A.B.N., E.N.; writing—original draft preparation/introduction K.-A.B.N., H.H., J.S.Y., E.N./method K.-A.B.N., H.H., E.N./results. E.N./discussion K.-A.B.N., E.N., H.H., J.S.Y.; writing—review and editing, K.-A.B.N., E.N., H.H., J.S.Y.; visualization, K.-A.B.N. and E.N.; project administration, K.-A.B.N.; funding acquisition, K.-A.B.N. All authors have read and agreed to the published version of the manuscript.

Funding: This research was funded by the Research Council of Norway, grant number 238030.

Institutional Review Board Statement: Regional Committee for Medical and Health Research Ethics Sør-øst. IRB ID: 19732.

Informed Consent Statement: Informed consent was obtained from all families involved in the study.

Data Availability Statement: The data are available in the services for sensitive data at the University of Oslo and can be obtained by contacting the first author.

Acknowledgments: We would like to thank the participating children and parents, schools, and research assistants as well as the funder of the study: grant number 238030 from the Research Council of Norway.

Conflicts of Interest: The authors declare no conflict of interest.

References

1. American Psychiatric Association. *Diagnostic and Statistical Manual of Mental Disorders*; APA: Washington, DC, USA, 2013.
2. Smith, A.; Weber, C. How stuttering develops: The multifactorial dynamic pathways theory. *J. Speech Lang. Hear. Res.* **2017**, *60*, 2483–2505. [CrossRef]
3. Tumanova, V.; Conture, E.G.; Lambert, E.W.; Walden, T.A. Speech disfluencies of preschool-age children who do and do not stutter. *J. Commun. Disord.* **2014**, *49*, 25–41. [CrossRef]
4. Evans, D. The development of language abilities in mongols: A correlational study. *J. Ment. Defic. Res.* **1977**, *21*, 103–117. [CrossRef]
5. Preus, A. Stuttering in Down's syndrome. *Scand. J. Educ. Res.* **1972**, *16*, 89–104. [CrossRef]
6. Van Borsel, J.; Vandermeulen, A. Cluttering in Down syndrome. *Folia Phoniatr. Logop.* **2008**, *60*, 312–317. [CrossRef]
7. Annaz, D.; Karmiloff-Smith, A.; Johnson, M.H.; Thomas, M.S.C. A cross-syndrome study of the development of holistic face recognition in children with autism, Down syndrome, and Williams syndrome. *J. Exp. Child Psychol.* **2009**, *102*, 456–486. [CrossRef] [PubMed]
8. Pinter, J.D.; Eliez, S.; Schmitt, J.E.; Capone, G.T.; Reiss, A.L. Neuroanatomy of Down's syndrome: A high-resolution MRI Study. *Am. J. Psychiatry* **2001**, *158*, 1659–1665. [CrossRef] [PubMed]
9. European Commission Home Page. Prevalence Charts and Tables. Available online: https://eu-rd-platform.jrc.ec.europa.eu/eurocat/eurocat-data/prevalence_en (accessed on 12 April 2021).
10. Chapman, R.S. Language development in children and adolescents with Down syndrome. In *Handbook of Child Language*; Fletcher, P., Mac-Whinney, B., Eds.; Blackwell: London, UK, 1995; pp. 641–663.
11. Næss, K.A.B.; Lyster, S.A.H.; Hulme, C.; Melby-Lervåg, M. Language and verbal short-term memory skills in children with Down syndrome: A meta-analytic review. *Res. Dev. Disabil.* **2011**, *32*, 2225–2234. [CrossRef]
12. Austeng, M.E.; Akre, H.; Falkenberg, E.S.; Øverland, B.; Abdelnoor, M.; Kværner, K.J. Hearing level in children with Down syndrome at the age of eight. *Res. Dev. Disabil.* **2013**, *34*, 2251–2256. [CrossRef]
13. Austeng, M.E.; Akre, H.; Øverland, B.; Abdelnoor, M.; Falkenberg, E.S.; Kværner, K.J. Otitis media with effusion in children with in Down syndrome. *Int. J. Pediatr. Otorhinolaryngol.* **2013**, *77*, 1329–1332. [CrossRef] [PubMed]
14. Nightengale, E.; Yoon, P.; Wolter-Warmerdam, K.; Daniels, D.; Hickey, F. Understanding hearing and hearing loss in children with Down syndrome. *Am. J. Audiol.* **2017**, *26*, 301–308. [CrossRef]
15. Bhagyalakshmi, G.; Renukarya, A.; Rajangam, S. Metric analysis of the hard palate in children with Down syndrome—A comparative study. *Down Syndr. Res. Pract.* **2007**, 55–59. [CrossRef]
16. Roberts, J.; Price, J.; Barnes, E.; Nelson, L.; Burchinal, M.; Hennon, E.A.; Moskowitz, L.; Edwards, A.; Malkin, C.; Anderson, K.; et al. Receptive vocabulary, expressive vocabulary, and speech production of boys with fragile X syndrome in comparison to boys with Down syndrome. *Am. J. Ment. Retard.* **2007**, *112*, 177–193. [CrossRef]
17. Suri, S.; Tompson, B.D.; Cornfoot, L. Cranial base, maxillary and mandibular morphology in Down syndrome. *Angle Orthod.* **2010**, *80*, 861–869. [CrossRef] [PubMed]
18. Pulina, F.; Vianello, R.; Lanfranchi, S. Cognitive profiles in individuals with Down syndrome. In *International Review of Research in Developmental Disabilities*; Lanfranchi, S., Ed.; Academic Press: Cambridge, MA, USA, 2019; Volume 56, pp. 67–92.
19. Jarrold, C.; Baddeley, A.D.; Phillips, C.E. Verbal short-term memory in Down syndrome: A problem of memory, audition, or speech? *J. Speech Lang. Hear. Res.* **2002**, *45*, 531–544. [CrossRef]
20. Tungate, A.S.; Conners, F.A. Executive function in Down syndrome: A meta-analysis. *Res. Dev. Disabil.* **2021**, *108*, 103802. [CrossRef] [PubMed]
21. Mason-Apps, E.; Stojanovik, V.; Houston-Price, C.; Seager, E.; Buckley, S. Do infants with Down syndrome show an early receptive language advantage? *J. Speech Lang. Hear. Res.* **2020**, *63*, 585–598. [CrossRef] [PubMed]
22. Caselli, M.C.; Monaco, L.; Trasciani, M.; Vicari, S. Language in Italian children with Down syndrome and with specific language impairment. *Neuropsychology* **2008**, *22*, 27–35. [CrossRef]
23. Price, J.R.; Roberts, J.E.; Hennon, E.A.; Berni, M.C.; Anderson, K.L.; Sideris, J. Syntactic complexity during conversation of boys with fragile X syndrome and Down syndrome. *J. Speech Lang. Hear. Res.* **2008**, *51*, 3–15. [CrossRef]

24. Næss, K.A.B.; Lervåg, A.; Lyster, S.A.H.; Hulme, C. Longitudinal relationships between language and verbal short-term memory skills in children with Down syndrome. *J. Exp. Child Psychol.* **2015**, *135*, 43–55. [CrossRef]
25. Laws, G.; Bishop, D.V. A comparison of language abilities in adolescents with Down syndrome and children with specific language impairment. *J. Speech Lang. Hear. Res.* **2003**, *46*, 1324–1339. [CrossRef]
26. Næss, K.A.B. Development of phonological awareness in Down syndrome: A meta-analysis and empirical study. *Dev. Psychol.* **2016**, *52*, 177–190. [CrossRef]
27. Eggers, K.; Van Eerdenbrugh, S. Speech disfluencies in children with Down syndrome. *J. Commun. Disord.* **2018**, *71*, 72–84. [CrossRef]
28. Dodd, B.; Thompson, L. Speech disorder in children with Down's syndrome. *J. Intellect. Disabil. Res.* **2001**, *45*, 308–316. [CrossRef] [PubMed]
29. Smith, E.; Næss, K.B.; Jarrold, C. Assessing pragmatic communication in children with Down syndrome. *J. Commun. Disord.* **2017**, *68*, 10–23. [CrossRef]
30. Næss, K.-A.B.; Ostad, J.; Nygaard, E. Differences and Similarities in Predictors of Expressive Vocabulary Development between Children with Down Syndrome and Young Typically Developing Children. *Brain Sci.* **2021**, *11*, 312. [CrossRef] [PubMed]
31. Witecy, B.; Penke, M. Language comprehension in children, adolescents, and adults with Down syndrome. *Res. Dev. Disabil.* **2017**, *62*, 184–196. [CrossRef] [PubMed]
32. Ntourou, K.; Conture, E.G.; Lipsey, M.W. Language abilities of children who stutter: A meta-analytical review. *Am. J. Speech-Lang. Pathol.* **2011**, *20*, 163–179. [CrossRef]
33. Anderson, J.D.; Conture, E.G. Language abilities of children who stutter: A preliminary study. *J. Fluen. Disord.* **2000**, *25*, 283–304. [CrossRef]
34. Anderson, J.D.; Pellowski, M.W.; Conture, E.G. Childhood stuttering and dissociations across linguistic domains. *J. Fluen. Disord.* **2005**, *30*, 219–253. [CrossRef]
35. Bloodstein, O.; Bernstein Ratner, N. *A Handbook on Stuttering*; Thomson-Delmar: Clifton Park, NY, USA, 2008.
36. Yairi, E.; Seery, C.H. *Stuttering: Foundations and Clinical Applications*, 2nd ed.; Pearson: Boston, MA, USA, 2015.
37. Johnson, W. Measurements of oral reading and speaking rate and disfluency of adult male and female stutterers and nonstutterers. *J. Speech Hear. Disord.* **1961**, *7*, 1–20.
38. Schieve, L.A.; Boulet, S.L.; Boyle, C.; Rasmussen, S.A.; Schendel, D. Health of children 3 to 17 years of age with Down syndrome in the 1997–2005 national health interview survey. *Pediatrics* **2009**, *123*, e253–e260. [CrossRef] [PubMed]
39. Van Riper, C. *The Nature of Stuttering*; Prentice-Hall, Inc.: Englewood Cliffs, NJ, USA, 1982.
40. Månsson, H. Childhood stuttering: Incidence and development. *J. Fluen. Disord.* **2000**, *25*, 47–57. [CrossRef]
41. Reilly, S.; Onslow, M.; Packman, A.; Cini, E.; Conway, L.; Ukoumunne, O.C.; Bavin, E.L.; Prior, M.; Eadie, P.; Block, S.; et al. Natural history of stuttering to 4 years of age: A prospective community-based study. *Pediatrics* **2013**, *132*, 460–467. [CrossRef]
42. Kent, R.D.; Vorperian, H.K. Speech impairment in Down syndrome: A review. *J. Speech Lang. Hear. Res.* **2013**, *56*, 178–210. [CrossRef]
43. Einarsdóttir, J.T.; Crowe, K.; Kristinsson, S.H.; Másdóttir, T. The recovery rate of early stuttering. *J. Fluen. Disord.* **2020**, *64*, 105764. [CrossRef] [PubMed]
44. Sjøstrand, Å.; Kefalianos, E.; Hofslundsengen, H.; Guttormsen, L.S.; Kirmess, M.; Lervåg, A.; Hulme, C.; Naess, K.A.B. Non-pharmacological interventions for stuttering in children aged between birth and six years. *Cochrane Database Syst. Rev.* **2019**. [CrossRef]
45. Jansson-Verkasalo, E.; Silvén, M.; Lehtiö, I.; Eggers, K. Speech disfluencies in typically developing Finnish-speaking children—Preliminary results. *Clin. Linguist. Phon.* **2020**, 1–20. [CrossRef]
46. Coppens-Hofman, M.C.; Terband, H.R.; Maassen, B.A.M.; van Schrojenstein Lantman-De Valk, H.M.J.; van Zaalen-op't Hof, Y.; Snik, A.F.M. Dysfluencies in the speech of adults with intellectual disabilities and reported speech difficulties. *J. Commun. Disord.* **2013**, *46*, 484–494. [CrossRef] [PubMed]
47. Harasym, J.; Langevin, M. Stuttering treatment for a school-age child with Down syndrome: A descriptive case report. *J. Fluen. Disord.* **2012**, *37*, 253–262. [CrossRef] [PubMed]
48. Wilcox, A. An investigation into non-fluency in Down's syndrome. *Br. J. Disord. Commun.* **1988**, *23*, 153–170. [CrossRef]
49. Zigman, W.B. Atypical aging in Down syndrome. *Dev. Disabil. Res. Rev.* **2013**, *18*, 51–67. [CrossRef]
50. Klimova, B.; Kuca, K. Speech and language impairments in dementia. *J. Appl. Biomed.* **2016**, *14*, 97–103. [CrossRef]
51. Verma, M.; Howard, R.J. Semantic memory and language dysfunction in early Alzheimer's disease: A review. *Int. J. Geriatr. Psychiatry* **2012**, *27*, 1209–1217. [CrossRef]
52. Pulsifer, M.B.; Evans, C.L.; Hom, C.; Krinsky-McHale, S.J.; Silverman, W.; Lai, F.; Lott, I.; Schupf, N.; Wen, J.; Rosas, H.D. Language skills as a predictor of cognitive decline in adults with Down syndrome. *Alzheimer's Dement. (Amst. Neth.)* **2020**, *12*, e12080. [CrossRef] [PubMed]
53. Salihović, N.; Hasanbašić, S.; Begic, L. Incidence of Stuttering in School-Age Children with Down Syndrome. *J. Spec. Educ. Rehabil.* **2012**, *13*. [CrossRef]
54. Craig, A.; Tran, Y. The Epidemiology of Stuttering: The Need for Reliable Estimates of Prevalence and Anxiety Levels over the Lifespan. *Int. J. Speech-Lang. Pathol.* **2009**, *7*, 41–46. [CrossRef]

55. Næss, K.A.B. Language and Reading Development in Children with Down Syndrome. Ph.D. Thesis, Reprosentralen, Oslo, Norway, 2012.
56. Wechsler, D. *WPPSI-III: Wechsler Preschool and Primary Scale of Intelligence*, 3rd ed.; Manual. (Swedish Version by Tidemann, E., 2005); The Psychological Corporation, Harcourt Assessment: San Antonio, TX, USA, 2002.
57. Dunn, L.M.; Dunn, L.M.; Whetton, C.; Burley, J. *British Picture Vocabulary Scale*; Nfer-Nelson: London, UK, 1997.
58. Lyster, S.A.H.; Horn, E.; Rygvold, A.L. Ordforråd og ordforrådsutvikling hos norske barn og unge. *Spesialpedagogikk* **2010**, *75*, 35–43.
59. Bishop, D.V.M. *Test for Reception of Grammar*; Version 2 (TROG-2) Manual; Harcourt Assessment: London, UK, 2003.
60. Lyster, S.A.H.; Horn, E. *Test for Reception of Grammar (TROG-2)*; Norwegian Version; Pearson Assessment: London, UK, 2009.
61. Gjessing, H.; Nygaard, H. *ITPA. Håndbok. Norsk Utgave [ITPA Manual, Norwegian Version]*; Universitetsforlaget: Oslo, Norway, 1975.
62. Kirk, S.A.; McCarthy, J.; Kirk, W.D. *The Illinois Test of Psycholinguistic Abilities*; University of Illinois Press: Urbana, IL, USA, 1967.
63. Carroll, J.M.; Snowling, M.J.; Stevenson, J.; Hulme, C. The development of phonological awareness in preschool children. *Dev. Psychol.* **2003**, *39*, 913–923. [CrossRef]
64. Furnes, B.; Samuelsson, S. Preschool cognitive and language skills predicting kindergarten and grade 1 reading and spelling: A cross-linguistic comparison. *J. Res. Read.* **2009**, *32*, 275–292. [CrossRef]
65. Gathercole, S.E.; Baddeley, A. *Children's Test of Nonword Repetition*; Pearson Assessment: London, UK, 1996.
66. Child Language & Learning's Tests. *Tests Developed to a Longitudinal Study*; Department of Special Needs Education, University of Oslo: Oslo, Norway, 2007.
67. Landers, R.N. Computing intraclass correlations (ICC) as estimates of interrater reliability in SPSS. *Winnower* **2015**, *2*, e143518.181744. [CrossRef]
68. Althouse, A.D. Adjust for multiple comparisons? It's not that simple. *Ann. Thorac. Surg.* **2016**, *101*, 1644–1645. [CrossRef] [PubMed]
69. Ferguson, C.J. An effect size primer: A guide for clinicians and researchers. *Prof. Psychol. Res. Pract.* **2009**, *40*, 532–538. [CrossRef]
70. Hochberg, Y.; Benjamini, Y. More powerful procedures for multiple significance testing. *Stat. Med.* **1990**, *9*, 811–818. [CrossRef]
71. Yairi, E.; Ambrose, N. Epidemiology of stuttering: 21st century advances. *J. Fluen. Disord.* **2013**, *38*, 66–87. [CrossRef]
72. Kelman, E.; Nicholas, A. *Practical Intervention for Early Childhood Stammering: Palin PCI*; Speechmark Publishing Limited: Milton Keynes, UK, 2008.
73. Loveall, S.J.; Hawthorne, K.; Gaines, M. A meta-analysis of prosody in autism, Williams syndrome, and Down syndrome. *J. Commun. Disord.* **2021**, *89*, 106055. [CrossRef]
74. Jahromi, L.B.; Gulsrud, A.; Kasari, C. Emotional competence in children with Down Syndrome: Negativity and regulation. *Am. J. Ment. Retard.* **2008**, *113*, 32–43. [CrossRef]
75. Næss, K.-A.B.; Nygård, E.; Dolva, A.S.; Ostad, J.; Lyster, S. The profile of social functioning in children with Down syndrome. *Disabil. Rehabil.* **2016**, 1–12. [CrossRef]
76. Staróbole Juste, F.; de Andrade, C.R.F. Speech disfluency types of fluent and stuttering individuals: Age effects. *Folia Phoniatr. Logop.* **2011**, *63*, 57–64. [CrossRef] [PubMed]
77. Westby, C.E. Language performance of stuttering and nonstuttering children. *J. Commun. Disord.* **1974**, *12*, 133–145. [CrossRef]
78. Bernstein Ratner, N.; Silverman, S. Parental perceptions of children's communicative development at stuttering onset. *J. Speech Lang. Hear. Res.* **2000**, *43*, 1252–1263. [CrossRef] [PubMed]
79. Luckman, C.; Wagovich, S.A.; Weber, C.; Brown, B.; Chang, S.E.; Hall, N.E.; Bernstein Ratner, N. Lexical diversity and lexical skills in children who stutter. *J. Fluen. Disord.* **2020**, *63*, 105747. [CrossRef]
80. Murray, H.L.; Reed, C.G. Language abilities of preschool stuttering children. *J. Fluen. Disord.* **1977**, *2*, 171–176. [CrossRef]
81. Ratner, N.; Sih, C. Effects of Gradual Increases in Sentence Length and Complexity on Children's Dysfluency. *J. Speech Hear. Disord.* **1987**, *52*, 278–287. [CrossRef] [PubMed]
82. Brundage, S.B.; Bernstein Ratner, N. Measurement of stuttering frequency in children's speech. *J. Fluen. Disord.* **1989**, *14*, 351–358. [CrossRef]
83. Coulter, C.E.; Anderson, J.D.; Conture, E.G. Childhood stuttering and dissociations across linguistic domains: A replication and extension. *J. Fluen. Disord.* **2009**, *34*, 257–278. [CrossRef] [PubMed]
84. Chon, H.; Sawyer, J.; Ambrose, N.G. Differences of articulation rate and utterance length in fluent and disfluent utterances of preschool children who stutter. *J. Commun. Disord.* **2012**, *45*, 455–467. [CrossRef]
85. Gaines, N.D.; Runyan, C.M.; Meyers, S.C. A comparison of young stutterers' fluent versus stuttered utterances on measures of length and complexity. *J. Speech Lang. Hear. Res.* **1991**, *34*, 37–42. [CrossRef]
86. Kadi-Hanifi, K.; Howell, P. Syntactic analysis of the spontaneous speech of normally fluent and stuttering children. *J. Fluen. Disord.* **1992**, *17*, 151–170. [CrossRef]
87. Logan, K.; Conture, E. Length, grammatical complexity, and rate differences in stuttered and fluent conversational utterances of children who stutter. *J. Fluen. Disord.* **1995**, *20*, 35–61. [CrossRef]
88. Melnick, K.S.; Conture, E.G. Relationship of length and grammatical complexity to the systematic and nonsystematic speech errors and stuttering of children who stutter. *J. Fluen. Disord.* **2000**, *25*, 21–45. [CrossRef]
89. Tornick, G.B.; Bloodstein, O. Stuttering and sentence length. *J. Speech Hear. Res.* **1976**, *19*, 651–654. [CrossRef] [PubMed]

90. Watson, J.B.; Byrd, C.T.; Carlo, E.J. Effects of length, complexity, and grammatical correctness on stuttering in Spanish-speaking preschool children. *Am. J. Speech-Lang. Pathol.* **2011**, *20*, 209–220. [CrossRef]
91. Weiss, A.L.; Zebrowski, P.M. Disfluencies in the conversations of young children who stutter: Some answers about questions. *J. Speech Lang. Hear. Res.* **1992**, *35*, 1230–1238. [CrossRef]
92. Yaruss, J.S. Utterance length, syntactic complexity and childhood stuttering. *J. Speech Lang. Hear. Res.* **1999**, *42*, 329–344. [CrossRef] [PubMed]
93. Yaruss, J.S.; Newman, R.M.; Flora, T. Language and disfluency in nonstuttering children's conversational speech. *J. Fluen. Disord.* **1999**, *24*, 185–207. [CrossRef]
94. Buhr, A.P.; Jones, R.M.; Conture, E.G.; Kelly, E.M. The function of repeating: The relation between word class and repetition type in developmental stuttering. *Int. J. Lang. Commun. Disord.* **2016**, *51*, 128–136. [CrossRef] [PubMed]
95. Hakim, H.B.; Ratner, N.B. Nonword repetition abilities of children who stutter: An exploratory study. *J. Fluen. Disord.* **2004**, *29*, 179–199. [CrossRef]
96. Schaner-Wolles, C. Sprachentwicklung Bei Geistiger Behindering: Williams-Beuren-Syndrom und Down-Syndrom. In *Enzyklopädie der Psychologie: Spracherwerb*; Grimm, H., Ed.; Hogrefe Verlag für Psychologie: Göttingen, Germany, 2000; pp. 663–685.
97. Buckley, S.J.; Bird, G. *Speech and Language Development for Infants with Down Syndrome (0–5 Years)*; Down Syndrome Educational Trust: Southsea, UK, 2001.
98. Andreou, G.; Chartomatsidou, E. A review paper on the syntactic abilities of individuals with down syndrome. *Open J. Mod. Linguist.* **2020**, *10*, 480–523. [CrossRef]
99. van Bysterveldt, A.K.; Westerveld, M.F.; Gillon, G.; Foster-Cohen, S. Personal narrative skills of school-aged children with Down syndrome. *Int. J. Lang. Commun. Disord.* **2012**, *47*, 95–105. [CrossRef]
100. Arias-Trejo, N.; Barrón-Martínez, J.B. Language skills in down syndrome. In *Language Development and Disorders in Spanish-speaking Children*; Auza Benavides, A., Schwartz, R.G., Eds.; Springer International Publishing: Cham, Switzerland, 2017; pp. 329–341.
101. Laws, G.; Gunn, D. Relationships between reading, phonological skills and language development in individuals with Down syndrome: A five year follow-up study. *Read. Writ.* **2002**, *15*, 527–548. [CrossRef]
102. Nippold, M.A. Language development in children who stutter: A review of recent research. *Int. J. Speech-Lang. Pathol.* **2019**, *21*, 368–376. [CrossRef] [PubMed]
103. Silverman, E.M.; Williams, D.E. A comparison of stuttering and nonstuttering children in terms of five measures of oral language development. *J. Commun. Disord.* **1967**, *1*, 305–309. [CrossRef]
104. Sjøstrand, Å.; Guttormsen, L.S.; Melle, A.H.; Hoff, K.; Hansen, E.H.; Næss, K.-A.B. Treatment for stuttering in preschool aged children–A qualitative content analysis of existing treatment programs. In progress.
105. Hollister, J.; Alpermann, A.; Zebrowski, P. Dissociations within and across linguistic and motor domains in children who stutter. In *Proceedings of Annual Meeting of the American Speech-Language-Hearing Association*; ASHA: Atlanta, GA, USA, 2012.
106. Bray, M. Speech Production in People with Down Syndrome. Available online: https://www.researchgate.net/profile/Monica_Bray/publication/237406961_Speech_production_in_people_with_Down_syndrome/links/54c6712c0cf2911c7a58d93a.pdf (accessed on 18 April 2021).
107. Buckley, S.J. Language development in children with Down syndrome-reasons for optimism. *Down Syndr. Res. Pract.* **1993**, *1*, 3–9. [CrossRef]
108. Cleland, J.; Wood, S.; Hardcastle, W.; Wishart, J.; Timmins, C. Relationship between speech, oromotor, language and cognitive abilities in children with Down's syndrome. *Int. J. Lang. Commun. Disord.* **2010**, *45*, 83–95. [CrossRef] [PubMed]
109. Corrales-Astorgano, M.; Escudero-Mancebo, D.; González-Ferreras, C. Acoustic characterization and perceptual analysis of the relative importance of prosody in speech of people with Down syndrome. *Speech Commun.* **2018**, *99*, 90–100. [CrossRef]
110. Kumin, L. Intelligibility of speech in children with Down syndrome in natural settings: Parents' perspective. *Percept. Mot. Ski.* **1994**, *78*, 307–313. [CrossRef] [PubMed]

Article

Occurrence of Reading Skills in a National Age Cohort of Norwegian Children with Down Syndrome: What Characterizes Those Who Develop Early Reading Skills?

Kari-Anne B. Næss [1,*], Egil Nygaard [2] and Elizabeth Smith [3]

1 Department of Special Needs Education, University of Oslo, 0318 Oslo, Norway
2 Department of Psychology, University of Oslo, 0317 Oslo, Norway; egilny@psykologi.uio.no
3 School of Psychology, University of Cardiff, Cardiff CF10 3AS, UK; SmithL57@cardiff.ac.uk
* Correspondence: k.a.b.nass@uv.uio.no; Tel.: +47-922-40-741

Abstract: Children with Down syndrome are at risk of reading difficulties. Reading skills are crucial for social and academic development, and thus, understanding the nature of reading in this clinical group is important. This longitudinal study investigated the occurrence of reading skills in a Norwegian national age cohort of 43 children with Down syndrome from the beginning of first grade to third grade. Data were collected to determine which characteristics distinguished those who developed early reading skills from those who did not. The children's decoding skills, phonological awareness, nonverbal mental ability, vocabulary, verbal short-term memory, letter knowledge and rapid automatized naming (RAN) performance were measured annually. The results showed that 18.6% of the children developed early decoding skills by third grade. Prior to onset, children who developed decoding skills had a significantly superior vocabulary and letter knowledge than non-readers after controlling for nonverbal mental abilities. These findings indicate that early specific training that focuses on vocabulary and knowledge of words and letters may be particularly effective in promoting reading onset in children with Down syndrome.

Keywords: trisomy 21; decoding; vocabulary; letters; phonological awareness

Citation: Næss, K.-A.B.; Nygaard, E.; Smith, E. Occurrence of Reading Skills in a National Age Cohort of Norwegian Children with Down Syndrome: What Characterizes Those Who Develop Early Reading Skills? *Brain Sci.* **2021**, *11*, 527. https://doi.org/10.3390/brainsci11050527

Academic Editor: Margaret B. Pulsifer

Received: 8 March 2021
Accepted: 16 April 2021
Published: 21 April 2021

Publisher's Note: MDPI stays neutral with regard to jurisdictional claims in published maps and institutional affiliations.

1. Introduction

Reading is the cognitive process of decoding words and obtaining meaning from text [1]. Functional reading skills expand opportunities for learning and participation in educational and social activities at home, in school, at work and in society [2,3], substantially impacting individuals' lives. Due to their reduced cognitive capacity, children with Down syndrome often experience reading difficulties [4]. In recent decades, however, reading skills and other academic achievements have improved among children with Down syndrome [5]. These improvements could be due to the higher educational goals being set for them and their increased inclusion in mainstream education, where literacy is a key area of the curriculum [5–7]. However, the proportion who successfully develop reading skills in their early school years is unknown, and which early cognitive skills promote early reading success is unclear.

Understanding the development of functional reading skills in children with Down syndrome in the lead-up to entering primary school is crucial for planning early educational interventions. Therefore, this study aims to investigate the occurrence of reading skills in children with Down syndrome and to provide insight into early abilities prior to reading onset and formal education, as these abilities may differ between young readers and non-readers with Down syndrome.

1.1. Development of Reading

For beginning readers, much energy is dedicated to the technical part of the reading process to learn to decode; however, as reading becomes more fluent, children dedicate more energy to linguistic comprehension (cf. automaticity theory; [8,9]). In stage models describing technical development, children begin with a logographical strategy (sight word reading without knowing the alphabetical principle), followed by a phonological decoding strategy (phoneme–grapheme correspondence is gradually established, and phonemes are synthesized into syllables and words) and then an orthographic decoding strategy (the orthographic, phonetic and semantic identities of words are stored in long-term memory and can be directly accessed). Over time, increased print experience results in increased automaticity and fluent decoding (for an overview, see Frith [10]). Typically developing children become fluent readers at approximately 3rd grade [11].

Different stage models exist (see also, e.g., [12]), and they have been criticized for oversimplifying the process of decoding development (e.g., [13]), overlooking, for example, word familiarity, word complexity, whether a word appears alone or in context [14] and the transparency of language [15]. However, such models provide a general framework for understanding how children transition from one decoding strategy to the next (e.g., [16]), and they emphasize that there are different subskills of word identification [17].

1.2. Development of Reading in Children with Down Syndrome

It has been suggested that rather than progressing through the above stages in decoding development, children with Down syndrome tend to rely on logographic strategies throughout their school years and beyond [18]. Reliance on such strategies may explain the difficulties with non-word decoding observed in groups with Down syndrome [19,20]. However, some children with Down syndrome develop exceptional reading skills [21], and a small proportion perform in line with typically developing peers of the same chronological age [19]. Thus, in individuals with Down syndrome, poor decoding skills are certainly not inevitable.

1.3. The Occurrence of Reading in Children with Down Syndrome

Differences in school placement, access to intervention, how reading is taught and expectations about children's potential may lead to very different reading outcomes among children with Down syndrome. Thus, as Groen et al. [21] note, it is difficult to know what level of reading ability to expect at a given point in time. Additionally, longitudinal research investigating the occurrence of reading skills in children with Down syndrome is limited. However, a five-year longitudinal study by Bird et al. [22] and a two-year longitudinal study by Byrne et al. [23] both included children below the age of 13 years (at the beginning of the study) and reported that 83.3% and 87.5% of participants, respectively, were able to decode words at a measurable level by the end of the study. In a five-year longitudinal study by Laws and Gunn [24], who included a mix of children, adolescents and adults (10–24 years of age at the beginning of the study), there was an occurrence of 53% at the end of the study. Thus, in previous research, the occurrence of reading skills in individuals with Down syndrome varies widely. Notably, there are no cohort studies that provide age-specific occurrence data on the reading ability of children with Down syndrome. The two studies above focusing specifically on the occurrence of reading in children [22,23] include age ranges of 5 years and 8 years, age ranges within which children could be expected to show vast differences in ability [11]. The combination of the wide age ranges and small sample sizes ($n = 12$ [22]; $n = 24$ [23]; $n = 30$ [24]) in the abovementioned studies means that it is not possible to meaningfully break down occurrence by age. These limitations and the lack of cohort studies in the early years mean that we do not know what to expect with regard to children with Down syndrome at specific ages/stages of development, such as the early school years. In an age cohort, the reading abilities across children may be expected to be within a more confined range due to less age- and experience-related variance.

1.4. Variables Related to the Development of Reading Skills

To suggest appropriate intervention routes specifically adapted to the phenotype of Down syndrome, it is also critical to explore the variables related to early reading skills in this population. In typically developing children, there is a consensus that phonological skills play a key role; phonological awareness, verbal memory, rapid automatized naming (RAN) and/or letter knowledge have been repeatedly found to predict reading performance [25–27]. Notably, the development of phonological awareness has been found to be tied to basic lexical knowledge. Walley et al. [28] suggest that vocabulary growth leads to segmental lexical representations, which are thought to be important for explicit phonemic segmentation and phonemic awareness. Therefore, vocabulary has also been found to positively affect children's reading development (cf. [29,30]) and to differentiate between typical readers and poor readers [31,32]. The associations among phonological skills, vocabulary and reading may be logical since the decoding process proceeds through the previously presented stages: from visually driven coding between printed letters and word pronunciations to the more sophisticated use of phonological and lexical information aggregating more effective word recognition processes [10,33].

The role of phonological variables in predicting reading skills in children with Down syndrome has been debated. There is a consensus that children with Down syndrome generally have weak phonological skills (letter knowledge [19]; phonological awareness [34]; memory [35]), which may in itself call into question the impact of these skills on reading development and suggest that other variables may have stronger compensatory influences. However, the association between phonological skills and reading skills among children with Down syndrome varies across studies. While some studies have concluded that phonological variables (e.g., phonological awareness [18,36–40], memory [23,41,42], RAN [21], and letter knowledge [42]) play a key role in decoding outcomes, reading has also been observed in this population in the absence of certain phonological skills, e.g., phonological awareness [43]. Notably, in the existing longitudinal studies on reading development in children with Down syndrome, recruitment was conducted after reading onset. Since the relationship between phonological awareness and reading is suggested to be reciprocal in nature, phonological awareness may have promoted early reading, which in turn augmented the development of phonological awareness [44,45]. Therefore, variations in children's reading experiences may be associated with variations in both the level of mastery of phonological awareness and the strength of its association with reading.

Various studies indicate that language skills, including vocabulary, also play a role in the reading development of individuals with Down syndrome (e.g., [22,24,40]). Notably, Hulme et al. [19] and Boudreau [46] found that language was a stronger predictor of reading ability in children with Down syndrome than in nonverbal, mental-age-matched, typically developing children. For groups with an impaired phonological pathway and weak decoding skills, such as children with Down syndrome [39], semantic word knowledge has been argued to be more important [47,48]. Similarly, familiarity with the spoken form of a new word may be particularly helpful in supporting reading in these children, potentially providing some compensation for these other difficulties.

Finally, since there appears to be a weak but consistent relationship between nonverbal mental ability and general reading skills in typically developing children [49], nonverbal mental ability is also an important variable to consider, as children with Down syndrome usually have intellectual disabilities. Several studies have found indications of such an association. For example, Laws and Gunn [24] found evidence of a significantly higher nonverbal mental ability score in readers than in non-readers with Down syndrome (e.g., [24]). However, because of the low number of participants usually included in studies of children with Down syndrome, the unique contribution of phonological skills over and above nonverbal mental ability has seldom been reported.

In addition to the underlying cognitive skills mentioned above, other variables, such as the home literacy environment [50], socioeconomic status [51] and schooling [34], may impact the reading skills of those with Down syndrome. The effect of hearing on reading development in children with Down syndrome has also been debated (e.g., [19]). However, the present study focuses on understanding which cognitive variables are underlying strengths in children with Down syndrome who develop early reading skills; the findings may indicate which variables enhance reading ability in this population early in development. Supporting reading-associated variables from an early age could provide greater potential for future reading success; the effect of such support is a well-established finding in typically developing children (e.g., see the article on "Matthew effects" by Stanovich [52]) and may also apply to individuals with Down syndrome.

1.5. Summary and Research Questions

The occurrence of reading skills in children with Down syndrome across the early school years from the time when formal teaching starts remains unknown. Thus, few studies have provided guidelines on appropriate expectations with regard to reading outcomes and approaches to teaching reading to children with Down syndrome. In all of the abovementioned studies on occurrence, the majority of subjects varied widely in age and were recruited post-reading onset. Thus, the occurrence of reading skills in young children with Down syndrome internationally may not be as high as suggested in the previous literature. If this is the case, it is important to know, as it may influence, e.g., parents' views on their children's development and teachers' expectations.

Furthermore, knowledge of what differentiates those with early decoding skills from those without these skills is limited because the low number of participants in previous longitudinal studies has limited the number of predictive variables included in the analysis.

Therefore, in the present study, we ask the following research questions: (1) What is the occurrence of reading skills in a national age cohort of Norwegian children with Down syndrome in grades 1, 2 and 3 (ages 6–8), and (2) what distinguishes the cognitive profiles of readers and non-readers prior to reading onset? The present study focuses on an age cohort to provide more specific information about occurrence in relation to age to better inform expectations, and it also includes a measure of parental report alongside a standardized measure to draw comparisons between outcomes for these respective measures.

2. Materials and Methods

The results reported in this paper are original reading data obtained from a larger research project studying a national age cohort of 43 children with Down syndrome. The title of the project is "Language and reading development in children with Down syndrome" [53].

2.1. Participants

A Norwegian national age cohort of six-year-old children with Down syndrome (including every registered child across the country) was invited to participate; all habilitation services in Norway forwarded an informational letter and a consent form to the registered parents of each child with Down syndrome. The letter and consent form were approved in advance by the Regional Committees for Medical and Health Research Ethics. The families of forty-three children with Down syndrome accepted the invitation on the children's behalf (22 boys and 21 girls; chronological age: mean (M) = 75.78 months, SD = 3.48 months; nonverbal mental ability raw score (Block Design): M = 12.23, SD = 5.40). The families who accepted the invitation returned the consent form to the principal investigator. In addition to being 6 years old at the start of the study, the inclusion criteria were that the child did not have a comorbid diagnosis of autism spectrum disorder (ASD) and that Norwegian was the first language.

Among the readers, all children had trisomy 21 except one who had translocation. The non-readers showed almost identical percentages. All except two children had trisomy 21. One of the two had translocation, and the other had mosaic. All participants in both groups except for one of the non-readers went to ordinary primary school. At T1, there were quite similar occurrences between the two groups in regard to permanent hearing disability (25% vs. 30% for non-readers and readers, respectively, odds ratio (OR) = 0.77, 95% confidence interval (CI) 0.13–4.48, p = 0.77) and mean parental education (M = 2.51, SD = 1.07 vs. M = 2.56, SD = 1.02 for non-readers and readers, respectively, OR = 1.05, 95% CI 0.50–2.20, p = 0.91). However, at T1, the nonverbal mental ability of readers (M = 3.13, SD = 2.10) was substantially better than that of non-readers (M = 1.49, SD = 1.07, OR = 1.95, 95% CI 1.13–3.41, p = 0.01) (scaled scores based on the Block Design; for a description of the measure, see Section 2.3.3).

2.2. Data Collection

Data were collected through clinical assessment of the children and through parental questionnaires. The children were assessed every autumn for their first three school years. They were assessed individually in separate rooms in three sessions, typically on consecutive days. All answers were registered manually in the standardized test protocol, and expressive answers were audio recorded for subsequent verification. The parental questionnaire was sent to one parent of each participating child. Up to two reminders were sent if no answers were received by the deadline. The answers were automatically coded in SPSS.

2.3. Measures

Standardized procedures for the implementation and scoring of the tests were followed. The tests that were used were originally developed for typically developing children. Although they have not been specifically validated for children with Down syndrome, they have been commonly used in research involving this group of children. Internal consistency, which is a function of the number of test items and the average inter-correlation among the items for the current sample, exhibited reasonably good reliability for all tests (ranging between α = 0.77 and α = 0.95), except for the initial syllable measure (α = 0.57). Notably, for RAN, reliability was calculated by the intraclass correlation between RAN1 and RAN2 and was found to be moderate, ICC = 0.57, when using a two-way mixed-effects model with absolute agreement based on an average of the two measures.

2.3.1. Reading Measures

The dependent variable was children's reading skills, which were assessed using a standardized test for decoding and spelling, STAS-OA-1 [54]. In STAS-OA-1, children are shown a list of high-frequency, phonetically regular single words (without any visual context/support), and they are instructed to read the words aloud for 40 s. The word list starts with two-syllable words, and the length and difficulty of the words gradually increase. Children earn one point for every word read correctly. Spelling is not considered reading; for example, if a child says the letters "c", "a" and "t" separately, they do not score a point. Children have to synthesize the phonemes into a word to score a point. In this standardized measure, reading reflects the decoding of different word classes without any contextual support, which is essential for effective independent reading. The STAS is used in Norway as a standard reading assessment strategy for all children in mainstream schools, and it has been shown to be highly reliable (e.g., [55]). Children who scored more than 1 point on the test at any time point (first grade: T1, second grade: T2 or third grade: T3) were classified as "readers". This definition is based on earlier studies (e.g., [24]).

At T1, reading was also measured as part of a large digital parental questionnaire on different background measures categorizing the number of words the child could recognize based on the following scale: 0 words = 0; 1–5 words = 1; 6–10 words = 2; 11–15 words = 3; more than 15 words = 4. There was also a category for unknown.

2.3.2. Background Measures

In the background questionnaire, we also collected information about the types of Down syndrome, school types, permanent hearing loss and the parents' highest educational level.

For the types of Down syndrome, the response options were 1 = I do not know, 2 = trisomy 21, 3 = translocation and 4 = mosaic. For school types, the response options were 1 = ordinary school, 2 = special school and 3 = other. For permanent hearing loss, the response options were 0 = no and 1 = yes. For the parents' highest educational level, the response options were elementary school = 0, high school (1–2 years) = 1, high school (3–4 years) = 2, university level up to 3 years = 3 and university level 4 years or more = 4. A mean parental score was calculated based on the average of the mother's and father's educational level.

2.3.3. Nonverbal Mental Ability

Nonverbal mental ability was assessed via the Block Design subtest of the third edition of the Wechsler Preschool and Primary Scale of Intelligence (WPPSI-III) [56]. In this subtest, a child is shown several building blocks put together in a pattern either with blocks (items 1–13) or via a picture (items 13–20). The child then has to copy the block arrangement. There are twenty items in total. Two points are scored each time the child correctly copies the block arrangement. For the first six items, the child is allowed two attempts, earning one point if they are correct on the second attempt rather than the first. Specified starting points and discontinuation rules were followed.

2.3.4. Vocabulary

The Norwegian versions of the British Picture Vocabulary Scale (BPVS-II; [57,58]) and Picture Naming (WPPSI-III; [56]) were used to assess vocabulary. For each item in the BPVS-II, a child is shown four pictures and is then asked to point to the picture corresponding to the word spoken aloud by the examiner. The test consists of 144 items, with specified starting points and discontinuation rules. The child earns one point for each correct answer.

The Picture Naming (WPPSI-III; [56]) task involves showing a child a set of single pictures, one item at a time, and then asking them to name the pictures. One point is scored for each correct answer, and no penalties are assessed for articulation errors. The test consists of 38 items, with specified starting points and discontinuation rules.

Vocabulary was calculated as the mean of the z-values of the raw scores on each of the two tests.

2.3.5. Verbal Short-Term Memory

Verbal short-term memory skills were measured via word span, non-word repetition and sentence memory.

In the word span task [59], a child hears a list of spoken words, and their task is to repeat the words in the correct order. The length of the word list gradually increases. The child earns one point for every list repeated correctly and is not penalized for systematic articulation errors. The test consists of 24 items. All the children started on word list 1 and continued until the discontinuation point was reached.

Non-word repetition was assessed using a Norwegian version of the Children's Test of Non-Word Repetition [60,61]. In each trial, a child hears a non-word, which they have to repeat. The non-words vary in length from two to five syllables. The child earns one point for every correct item and is not penalized for systematic articulation errors. The test consists of 28 items.

Sentence memory was assessed by the Sentence Repetition Test (WPPSI-R; [62]), in which a child listens to spoken sentences that they need to repeat. The child earns one point for every correct item and is not penalized for systematic articulation errors. The test consists of 21 items. All the children continued until a discontinuation point was reached.

Z-values based on the raw scores on the three tests were combined into a measure of mean verbal short-term memory.

2.3.6. Letter Knowledge

The letter sound test from the Aston Index [63] was used to assess letter knowledge. Twenty-four letters are included (c, w, x, z, and q are excluded). The letters are presented in six rows of four letters each. Each time the examiner points to a letter, the child's task is to decode that letter. The child earns one point for each correct answer, with both letter names and letter sounds accepted as answers. The results presented are z-values based on raw scores.

2.3.7. Rapid Automatized Naming

Two tasks, object RAN tasks for young Norwegian children [59], were used to assess RAN. All words included in the tasks were high-frequency words usually acquired at a very early age. In the first RAN task (RAN1), a child is given a sheet of paper showing black and white drawings of a sun, boat, mouse, door and bus. The five pictures are shown randomly in four rows with five items in each row. The child is asked to name each picture. The number that the child named correctly and the time that it took them to complete the task are recorded. In the second RAN task (RAN2), the child carries out the same task as in RAN1, but the pictures are of a light, ball, boy, house and car. The scoring scheme for RAN2 is the same as that for RAN1. The mean summary scores were calculated for the z-value of the total amount of time used on each of the two tasks.

2.3.8. Phonological Awareness

Four implicit measures of phonological awareness (initial syllable matching, final syllable matching, rhyme matching and initial phoneme matching) and their standardized procedures were adapted from Carroll et al. [64]. In each of the tasks, a child is shown a puppet, such as "Frode the frog"; for the initial syllable matching test, the child is told that the puppet likes to collect words that start with the same syllable. For each item, the frog puppet holds a picture card in front of the child, while two more picture cards are laid on the table. The child is asked to point to the picture beginning with the same syllable. The task was presented in the same way for the other three phonological awareness measures (with a different puppet for each), where children were told that the puppet would like to collect words with the same final syllable, words that rhymed/sounded the same, or words with the same initial sound. For every correct answer across these four measures, the child earns one point, and the summary scores for each of the measures are calculated separately. Both of the syllable measures consist of 8 items, while the remaining two measures consist of 16 items each. The four test results were standardized (z-values) and combined into a mean phonological awareness score.

For all the measures, standardized procedures were followed. Practice examples were provided before the tests started to ensure that the children understood each task.

2.4. Analysis

The dependent variable, the STAS-OA1 test, was dichotomized due to skewed results. For the other measures, we have specified in the description above how each of the measures were combined into summary scores. Binary logistic regression analyses, both bivariate and controlled for nonverbal mental ability, were used to analyze the differences between groups. The assumptions for logistic regression analyses were satisfied based on Box–Tidwell tests and the variance inflation factors. There were no missing data on the STAS-OA1 or any of the predictors.

IBM SPSS version 27 was used for all the analyses, all the tests were two-tailed, and we used a significance level of 0.05. ORs greater than 2.74 were considered to indicate a medium effect; ORs greater than 4.72 were considered to indicate a large effect [65].

3. Results

3.1. Occurrence of Reading Skills in Children with Down Syndrome Aged 6–8

At T1, the year the children started school, none of the participants were characterized as a reader based on the standardized decoding measure. However, parental data showed that 81.4% of the children recognized at least some written words at this age. An overview of the number of words the children recognized based on parental reports is presented in Table 1.

Table 1. Overview of the number of words recognized by the children at T1 based on parental reports.

Number of Words the Child Can Read	N	Percentage
0 words	8	18.6%
1–5 words	20	46.5%
6–10 words	7	16.3%
11–15 words	2	4.7%
More than 15	3	7%
Unknown *	3	7%
Total	43	100%

* A parent reported that their child could recognize words, but the response to frequency was omitted.

The words recognized were mainly the children's own name, "mum" and "dad" or their family members' first names.

At T2, after one year of school, the proportion of readers was 11.6% on the STAS-OA-1 (α = 0.86), and by 3rd grade (T3), it had increased to 18.6%.

As shown in Table 2, each reader read 2–10 words at T2, corresponding to the use of a logographic and/or phonological decoding strategy. At T3, the range had increased to 2–38 words. However, participant R3 was an outlier; this child's score of 38 was consistent with the normal range for 4th grade (M = 39, SD = 23), corresponding to the use of an orthographic decoding strategy.

Table 2. Number of words on the STAS-OA1 read by each of the individual participants who were classified as readers at T1, T2 and T3.

Readers	T1	T2	T3
R1	0	2	9
R2	0	0	4
R3	0	10	38
R4	0	0	2
R5	0	0	11
R6	0	5	13
R7	0	4	6
R8	0	2	3

Note: The values represent the *number* of words read at each time point; R = reader.

3.2. Differences in Cognitive Profiles between Readers and Non-Readers with Down Syndrome Prior to Reading Onset

The means and standard deviations for nonverbal mental ability, language and reading-related measures at T1 for both groups are shown in Table 3.

As shown in Table 3, the readers performed better than the non-readers on all the measures, with all but RAN showing medium-to-large effect sizes even after controlling for nonverbal mental ability. Differences in vocabulary and letter knowledge were significant after controlling for nonverbal mental ability. Additionally, there were significant bivariate group differences in nonverbal mental ability, short-term memory and phonological awareness. However, the group differences in RAN were not significant.

Table 3. Mean standardized (*Z*) scores (*SD*) and *OR* for the cognitive measures at T1 of readers vs. non-readers at T3.

	Readers (*n* = 8)	Non-Readers (*n* = 35)	Bivariate Analyses		Controlled for Nonverbal Mental Ability	
	M (SD)	*M (SD)*	*OR* (95% *CI*)	*p*-Value	*OR* (95% *CI*)	*p*-Value
Nonverbal mental ability	0.65 (1.14)	-0.15 (0.92)	2.85 (1.00–8.19)	0.05		
Vocabulary	0.95 (0.50)	−0.22 (0.85)	9.86 (1.88–51.75)	0.007	8.04 (1.46–44.26)	0.02
Verbal short-term memory	0.54 (0.79)	−0.12 (0.61)	3.94 (1.20–12.96)	0.02	3.01 (0.82–11.05)	0.10
Letter knowledge	1.01 (1.16)	−0.23 (0.81)	3.07 (1.38–6.84)	0.006	2.77 (1.22–6.29)	0.02
RAN	−0.29 (0.23)	0.07 (0.91)	0.53 (0.17–1.64)	0.27	0.33 (0.08–1.42)	0.14
Phonological awareness	0.56 (0.50)	−0.13 (0.85)	4.75 (0.97–23.26)	0.05	5.83 (0.77–44.09)	0.09

Note: The odds ratio (*OR*) with the 95% confidence interval (*CI*) is from logistic regression analyses. Predictors were standardized (*Z*) before being entered into the models. Higher scores reflect better performance on all the measures, except for the RAN task, in which lower scores reflect faster naming speed. Bold = significant group differences ($p \leq 0.05$).

4. Discussion

This longitudinal study aimed to investigate the occurrence of reading skills in a Norwegian age cohort of children with Down syndrome at 1st (T1), 2nd (T2) and 3rd (T3) grade and to provide insight into early abilities, as these abilities may differ between children who do and do not develop early reading skills during this period. In particular, we were interested in which abilities are strengths in early readers prior to their reading onset. The data showed that the occurrence of reading skills was low but increased over the years. Vocabulary and letter knowledge were stronger in readers than in non-readers prior to their reading onset.

4.1. Occurrence of Reading Skills in a Norwegian Age Cohort of Children with Down Syndrome

According to the parental measure, the majority of the children were able to recognize some words at T1. None of the children could decode words on the standardized reading measure at this time point. The fact that children could mainly recognize names or a very limited set of words and were not able to decode words on a standardized decoding test might reflect that these children utilized a logographic strategy rather than having reached a phonological decoding level at this point in time. However, the occurrence of decoding increased over the years. By 2nd grade, 11.6% of the children with Down syndrome achieved measurable levels of decoding skills on the standardized measure, and this proportion increased to 18.6% of the sample by 3rd grade. This result may appear to be a low occurrence of reading skills compared to that of age-spread samples of individuals with Down syndrome from previous research (e.g., [22–24]). However, the apparent discrepancy in results may be associated with the following three factors. (1) The first is the methodological aspect of the present study. The inclusion of an entire national age cohort allowed for a relatively large sample (the largest possible *n* for this age group at a national level within the time frame of the current study period) and meant that all the children were the same age. As a result, no children entered the study with decoding skills on the standardized decoding measure; they all started to receive instruction in reading at the start of the study and received this instruction for a similar length of time in the study period. Apart from the requirement of being six years of age, having Norwegian as the first language and having no comorbidity of ASD, there were no selections made in the recruitment procedure, for example, no requirement of verbal skills in the children, no specifications regarding the area of the country, and no consideration of whether the children accessed specific support services. (2) The second is related to the Norwegian educational system. For example, the strong role of play in kindergarten in Norway [66] made it likely that systematic instruction in reading would not have been introduced to our study participants before they started school at six years of age. Due to the developmental profile of children with Down syndrome it usually takes longer to learn new skills compared to typically developing children (cf. [53]). It is therefore likely that the occurrence will gradually increase with age and length of training. This reasoning is also supported by our data since the occurrence of reading skills increased over the years. Similarly, previous longitudinal research on children with Down syndrome supports this reasoning; Laws and Gunn [24] found a large increase in the occurrence of reading skills from age 11 (33% of their participants) to age 16 (53% of their participants), which corresponds to an average increase of 4% per year. Assuming that the occurrence of reading continues at the same pace for each year children receive instruction in reading, our results after two years of school education can be considered relatively consistent with those of studies such as Laws and Gunn [24]. (3) The third is the available educational resources adapted for Norwegian students with Down syndrome. Norwegian is an infrequently used language with relatively few available materials and seminars for parents and teachers working specifically with children with Down syndrome, while previous studies on occurrence have usually been conducted in English-speaking countries (e.g., [22–24]), where Sue Buckley and her team made available reading materials for this group of children from a very early age along with seminars for their parents and teachers (Down Syndrome Education International https://www.down-syndrome.org/ (accessed on 18 April 2021)). Since the occurrence of decoding also varies greatly across previous research, the results of the present study complement earlier findings, applying data from school starters in a non-English-speaking country and using different inclusion criteria, measures and methodological approaches.

4.2. What Distinguishes Those Who Develop Early Decoding Skills from Those Who Do Not?

Given that individuals with Down syndrome tend to experience significant learning difficulties and considering the young age of the cohort of this study, it is not surprising that many children with Down syndrome are somewhat delayed in achieving phonological decoding skills compared to what is expected of their typically developing peers. However, our results indicate that there were significant differences in nonverbal mental ability between readers and non-readers, with those who developed early reading skills showing better nonverbal mental functioning. These results are in line with what is suggested both for typically developing children (c.f. [49]) and in a previous study of children with Down syndrome [24]. However, when the children's nonverbal mental ability was controlled for, there were other variables that accounted for the differences between readers and non-readers prior to their reading onset. These findings demonstrate that the early reading ability of children with Down syndrome was not solely the result of stronger nonverbal mental ability. However, better nonverbal mental ability may have given these students access to reading interventions.

Prior to primary school education and reading onset, readers and non-readers displayed significant differences in vocabulary breadth and letter sound knowledge. The wide confidence interval of the vocabulary measure may limit the credibility of the odds ratio. However, both of these variables have been found to be reliable predictors of decoding in typically developing children (vocabulary [29]; letter sound knowledge [26]).

The importance of vocabulary is also consistent with earlier research on Down syndrome by Boudreau [46], Hulme et al. [19] and Steele et al. [67], who indicate that vocabulary is a stronger predictor of reading among these children than among typically developing children. As discussed by Hulme et al. [19], receptive vocabulary and expressive vocabulary tap into knowledge regarding both the phonological and semantic forms of words, which may help a child to both decode and develop contextual expectations about words to read in a concrete way. In line with this previous research, our findings indicate that early lexical knowledge may assist children with Down syndrome in obtaining decoding skills.

Moreover, we found that children with Down syndrome who exhibited word reading skills had greater letter knowledge than non-readers, as has also been observed in children with [68] and without Down syndrome (e.g., [69,70]). To understand the alphabetical principle and to use an analytic-based decoding approach, letter knowledge is necessary. As Muter et al. [69] have hypothesized, knowledge of the sounds of letters is also crucial for phonological decoding; that is, children understand that letter clusters represent phonemes. In addition, learning letter sounds provides a measure of paired visual–phonological associative learning that may correspond to the basic mechanism that is a fundamental component of learning to decode words [33]. The combination of good vocabulary and letter knowledge may help a reader to understand that words are made by letters, develop phonemic sensitivity [71] and predict words to read.

Finally, there is some indication that short-term memory and phonological awareness are underlying strengths in early readers with Down syndrome, as the effect sizes were substantial. However, due to lack of power when taking nonverbal mental abilities into account, the importance of these variables in children with Down syndrome learning to read are still inconclusive. Several previous studies have concluded that phonological awareness is reliably related to reading skills in children with Down syndrome (e.g., [36]); however, these studies measured the phonological measures post-reading onset. Based on the inconclusive results in the present study, we cannot interpret whether phonological skills are also important pre-reading onset. Thus, more studies are needed in future to clarify the role of phonological variables in the reading development of children with Down syndrome. It is also worth considering that reading development itself may promote phonological awareness, leading to differences in this outcome based on whether it is measured pre- versus post-reading onset [44,72].

4.3. Limitations and Strengths of the Study Design and Methods

This research is the first international longitudinal study investigating a relatively large age cohort of children with Down syndrome just starting school. In longitudinal studies, attrition usually occurs [73]. Nevertheless, in this study, no attrition occurred. We observed highly significant differences between readers and non-readers, which revealed clear early strengths among the readers. Future research is needed to determine whether each of these specific strengths of early readers plays a causal role in promoting reading development in children with Down syndrome. This study represents an important first step in identifying appropriate variables to investigate in future predictive studies.

Compared to other studies on reading skills in children with Down syndrome, the present study is robust in terms of its sample size; however, in regard to the power of the statistical analysis, the number of participants is still limited. To reduce the number of variables and possible bias of multiple comparisons or multicollinearity, we combined the scores of related predictors.

Children with Down syndrome may be slower at processing information than typically developing children, and therefore, the time frame for the standardized reading test may be challenging for them. However, we compared reading skills between two groups of children with Down syndrome, and all the subjects were likely to have the same processing problems. It could be argued that this reading test (the STAS test) underestimates the occurrence of reading skills in children with Down syndrome. However, this test is frequently used in Norwegian schools and is representative of how children's decoding skills are usually measured among their typically developing peers. Additionally, similar standardized subtests were used in previous studies on occurrence (e.g., the Kaufman Assessment Battery for Children [24] and the Woodcock Reading Mastery Test [22]). Based on the slow reading progress and the fact that a relatively high percentage of children scored 0 correct answers on the standardized measure but scored higher on the parental reports, this situation may call for more sensitive and reliable measures designed specifically for children with Down syndrome in the future. These future measures should take the reading development process into account to detect small changes in children's reading performance.

We did not collect specific data about reading interventions among the children, but the national curriculum for mainstream schools focuses heavily on phonological awareness and reading instruction for grades 1 and 2 (LK06). Where possible, all children, including those with Down syndrome, follow this curriculum.

5. Conclusions

This study demonstrates that children with Down syndrome can develop reading skills by 3rd grade. Our study extends the current literature by examining the age-specific occurrence of reading across the participants' first three years of school in Norway and by highlighting variables that may be underlying strengths of early readers with Down syndrome. Specifically, early readers exhibited significant strengths in vocabulary and letter knowledge skills prior to their reading onset, in addition to stronger nonverbal mental ability. As the children were pre-readers when these data were collected, the strengths cannot be attributed to learning via reading. Thus, these findings reinforce the need to consider children's language skills as well as their understanding of the alphabetical principle; early systematic training in vocabulary and play with letter knowledge could play a key role in promoting the development of early reading skills in children with Down syndrome.

Author Contributions: Conceptualization, K.-A.B.N. and E.S.; methodology, K.-A.B.N. and E.N.; formal analysis, E.N. and K.-A.B.N.; investigation, K.-A.B.N.; resources, K.-A.B.N.; data curation, K.-A.B.N.; writing—original draft preparation/introduction K.-A.B.N. and E.S./method K.-A.B.N. and E.N.; results E.N.; discussion K.-A.B.N. and E.S.; writing—review and editing, K.-A.B.N., E.S., and E.N.; visualization, K.-A.B.N. and E.N.; project administration, K.-A.B.N.; funding acquisition, K.-A.B.N. All authors have read and agreed to the published version of the manuscript.

Funding: This research was funded by the Research Council of Norway, grant number 238030.

Institutional Review Board Statement: Regional Committee for Medical and Health Research Ethics Sør-øst. IRB ID: 19732.

Informed Consent Statement: Informed consent was obtained from all families involved in the study.

Data Availability Statement: The data are available in the services for sensitive data at the University of Oslo and can be obtained by contacting the first author.

Acknowledgments: We would like to thank the participating children and parents, schools and research assistants as well as the funder of the study: grant number 238030 from the Research Council of Norway.

Conflicts of Interest: The authors declare no conflict of interest.

References

1. Gough, P.B.; Tunmer, W.E. Decoding, reading, and reading disability. *Remedial Spec. Educ.* **1986**, *7*, 6–10. [CrossRef]
2. Bigozzi, L.; Tarchi, C.; Vagnoli, L.; Valente, E.; Pinto, G. Reading fluency as a predictor of school outcomes across grades 4–9. *Front. Psychol.* **2017**, *8*, 200. [CrossRef] [PubMed]
3. Ritchie, S.J.; Bates, T.C. Enduring links from childhood mathematics and reading achievement to adult socioeconomic status. *Psychol. Sci.* **2013**, *24*, 1301–1308. [CrossRef] [PubMed]
4. Kendeou, P.; van den Broek, P.; Helder, A.; Karlsson, J. A cognitive view of reading comprehension: Implications for reading difficulties. *Learn. Disabil. Res. Pract.* **2014**, *29*, 10–16. [CrossRef]
5. Snowling, M.; Nash, H.; Henderson, L. The development of literacy skills in children with Down syndrome: Implications for intervention. *Down Syndr. Res. Pract.* **2008**, *12*, 62–66.
6. Beard, R. *National Literacy Strategy: Review of Research and Other Related Evidence*; Department for Education and Employment: London, UK, 1998.
7. Næss, K. *A.B. Language and Reading Skills in Children with Down Syndrome*; Repres;ntralen: Oslo, Norway, 2012.
8. LaBerge, D.; Samuels, S.J. Toward a theory of automatic information processing in reading. *Cogn. Psychol.* **1974**, *6*, 293–323. [CrossRef]
9. Samules, S.J.; Farstrup, A.E. *What Reasearch Has to Say about Reading Instruction*; International Reading Association Inc.: Newark, Delaware, 1992.

10. Frith, U. Beneath the surface of developmental dyslexia. In *Surface Dyslexia: Neurological and Cognitive Studies of Phonological Reading*; Patterson, K., Marshall, J., Coltheart, M., Eds.; Lawrence Erlbaum: Hillsdale, NJ, USA, 1985; pp. 301–330.
11. Snow, C.E.; Burns, M.S.; Griffin, P. *Preventing Reading Difficulties in Young Children*; National Academies Press: Washington, DC, USA, 1998.
12. Ehri, L.C. Phases of development in learning to read words by sight. *J. Res. Read.* **1995**, *18*, 116–125. [CrossRef]
13. Beech, J.R. Ehri's model of phases of learning to read: A brief critique. *J. Res. Read.* **2005**, *28*, 50–58. [CrossRef]
14. Ehri, L.C.; McCormick, S. Phases of word learning: Implications for instruction with delayed and disabled readers. *Read. Writ. Q.* **1998**, *14*, 135–163. [CrossRef]
15. Wimmer, H.; Hummer, P. How German-speaking first graders read and spell: Doubts on the importance of the logographic stage. *Appl. Psycholinguist.* **1990**, *11*, 349–368. [CrossRef]
16. Ellis, N. Interactions in the development of reading and spelling: Stages, strategies, and exchange of knowledge. In *Learning to Spell Research Theory and Practice across Languages*; Perfetti, A., Rieben, L., Fayol, M., Eds.; Routledge: Mahwah, NJ, USA, 1997; pp. 271–294.
17. Loveall, S.J.; Conners, F.A. Reading skills in Down syndrome: An examination of orthographic knowledge. *Am. J. Intellect. Dev. Disabil.* **2016**, *121*, 95–110. [CrossRef]
18. Roch, M.; Jarrold, C. A comparison between word and nonword reading in Down syndrome: The role of phonological awareness. *J. Commun. Disord.* **2008**, *41*, 305–318. [CrossRef]
19. Hulme, C.; Goetz, K.; Brigstocke, S.; Nash, H.M.; Lervåg, A.; Snowling, M.J. The growth of reading skills in children with Down syndrome. *Dev. Sci.* **2012**, *15*, 320–329. [CrossRef] [PubMed]
20. Næss, K.A.B.; Melby-Lervåg, M.; Hulme, C.; Lyster, S.A.H. Reading skills in children with Down syndrome: A meta-analytic review. *Res. Dev. Disabil.* **2012**, *33*, 737–747. [CrossRef]
21. Groen, M.A.; Laws, G.; Nation, K.; Bishop, D.V.M. A case of exceptional reading accuracy in a child with Down syndrome: Underlying skills and the relation to reading comprehension. *Cogn. Neuropsychol.* **2006**, *23*, 1190–1214. [CrossRef] [PubMed]
22. Bird, E.K.R.; Cleave, P.L.; McConnell, L. Reading and phonological awareness in children with Down syndrome: A longitudinal study. *Am. J. Speech Lang. Pathol.* **2000**, *9*, 319–330. [CrossRef]
23. Byrne, A.; MacDonald, J.; Buckley, S. Reading, language and memory skills: A comparative longitudinal study of children with down syndrome and their mainstream peers. *Br. J. Educ. Psychol.* **2002**, *72*, 513–529. [CrossRef]
24. Laws, G.; Gunn, D. Relationships between reading, phonological skills and language development in individuals with Down syndrome: A five year follow-up study. *Read. Writ.* **2002**, *15*, 527–548. [CrossRef]
25. Caravolas, M.; Lervåg, A.; Defior, S.; Seidlová Málková, G.; Hulme, C. Different patterns, but equivalent predictors, of growth in reading in consistent and inconsistent orthographies. *Psychol. Sci.* **2013**, *24*, 1398–1407. [CrossRef]
26. Clayton, F.J.; West, G.; Sears, C.; Hulme, C.; Lervåg, A. A longitudinal study of early reading development: Letter-sound knowledge, phoneme awareness and ran, but not letter-sound integration, predict variations in reading development. *Sci. Stud. Read.* **2020**, *24*, 91–107. [CrossRef]
27. Hulme, C.; Snowling, M.J. The interface between spoken and written language: Developmental disorders. *Philos. Trans. R Soc. Lond. B Biol. Sci.* **2014**, *369*, 20120395. [CrossRef]
28. Walley, A.C.; Metsala, J.L.; Garlock, V.M. Spoken vocabulary growth: Its role in the development of phoneme awareness and early reading ability. *Read. Writ.* **2003**, *16*, 5–20. [CrossRef]
29. Hagtvet, B.E.; Lyster, H.S.A.; Melbye-Lervåg, M.; Næss, B.K.A.; Hjetland, H.N.; Engevik, L.I.; Hølland, S.; Karlsen, J.; Klem, M.; Kruse, J. Ordforråd i førskolealder og senere leseferdigheter–En metaanalytisk tilnærming. *Spesialpedagogikk* **2011**, *1*, 34–49.
30. Yildirim, K.; Rasinski, T.; Ates, S.; Fitzgerald, S.; Zimmerman, B.; Yildiz, M. The relationship between reading fluency and vocabulary in fifth grade Turkish students. *Lit. Res. Instr.* **2014**, *53*, 72–89. [CrossRef]
31. Geva, E.; Massey-Garrison, A. A comparison of the language skills of ELLs and monolinguals who are poor decoders, poor comprehenders, or normal readers. *J. Learn. Disabil.* **2013**, *46*, 387–401. [CrossRef] [PubMed]
32. Snyder, L.S.; Downey, D.M. The language-reading relationship in normal and reading-disabled children. *J. Speech Hear. Res.* **1991**, *34*, 129–140. [CrossRef]
33. Hulme, C.; Snowling, M.J. Learning to read: What we know and what we need to understand better. *Child Dev. Perspect.* **2015**, *7*, 1–5. [CrossRef] [PubMed]
34. Næss, K.A.B. Development of phonological awareness in Down syndrome: A meta-analysis and empirical study. *Dev. Psychol.* **2016**, *52*, 177–190. [CrossRef]
35. Næss, K.A.; Lyster, S.A.; Hulme, C.; Melby-Lervåg, M. Language and verbal short-term memory skills in children with Down syndrome: A meta-analytic review. *Res. Dev. Disabil.* **2011**, *32*, 2225–2234. [CrossRef]
36. Lemons, C.J.; Fuchs, D. Phonological awareness of children with Down syndrome: Its role in learning to read and the effectiveness of related interventions. *Res. Dev. Disabil.* **2010**, *31*, 316–330. [CrossRef]
37. Cardoso-Martins, C.; Frith, U. Can individuals with Down syndrome acquire alphabetic literacy skills in the absence of phoneme awareness? *Read. Writ.* **2001**, *14*, 361–375. [CrossRef]
38. Fletcher, H.; Buckley, S. Phonological awareness in children with Down syndrome. *Down Syndr. Res. Pract.* **2002**, *8*, 11–18. [CrossRef] [PubMed]

39. Mengoni, S.E.; Nash, H.M.; Hulme, C. Learning to read new words in individuals with Down syndrome: Testing the role of phonological knowledge. *Res. Dev. Disabil.* **2014**, *35*, 1098–1109. [CrossRef]

40. Snowling, M.J.; Hulme, C.; Mercer, R.C. A deficit in rime awareness in children with Down syndrome. *Read. Writ.* **2002**, *15*, 471–495. [CrossRef]

41. Fowler, A.; Doherty, B. The basis of reading skill in young adults with Down syndrome. In *Down Syndrome Living and Learning in the Community*; Nadel, L., Rosenthal, D., Eds.; Wiley: New York, NY, USA, 1995; pp. 182–196.

42. Kennedy, E.; Flynn, M. Early phonological awareness and reading skills in children with Down syndrome. *Down Syndr. Res. Pract.* **2003**, *8*, 100–109. [CrossRef] [PubMed]

43. Cossu, G.; Rossini, F.; Marshall, J.C. When reading is acquired but phonemic awareness is not: A study of literacy in Down's syndrome. *Cognition* **1993**, *46*, 129–138. [CrossRef]

44. Perfetti, C.A.; Beck, I.; Bell, L.C.; Hughes, C. Phonemic knowledge and learning to read are reciprocal: A longitudinal study of first grade children. *Merrill-Palmer Q.* **1987**, *33*, 283–319.

45. Hogan, T.; Catts, H.W.; Little, T.D. The Relationship between Phonological Awareness and Reading: Implications for the Assessment of Phonological Awareness; 2005 Special Education and Communication Disorders Faculty Publications. Available online: https://digitalcommons.unl.edu/specedfacpub/14 (accessed on 1 March 2021).

46. Boudreau, D. Literacy skills in children and adolescents with Down syndrome. *Read. Writ.* **2002**, *15*, 497–525. [CrossRef]

47. McKay, A.; Davis, C.; Savage, G.; Castles, A. Semantic involvement in reading aloud: Evidence from a nonword training study. *J. Exp. Psychol. Learn. Mem. Cogn.* **2008**, *34*, 1495–1517. [CrossRef]

48. Taylor, J.S.H.; Plunkett, K.; Nation, K. The influence of consistency, frequency, and semantics on learning to read: An artificial orthography paradigm. *J. Exp. Psychol. Learn. Mem. Cogn.* **2011**, *37*, 60–76. [CrossRef]

49. Hayiou-Thomas, M.E.; Harlaar, N.; Dale, P.S.; Plomin, R. Genetic and environmental mediation of the prediction from preschool language and nonverbal ability to 7-year reading. *J. Res. Read.* **2006**, *29*, 50–74. [CrossRef]

50. van Bysterveldt, A.K.; Gillon, G.; Foster-Cohen, S. Literacy environments for children with down syndrome: What's happening at home? *Down Syndr. Res. Pract.* **2010**, *12*, 98–102.

51. Szumski, G.; Karwowski, M. School achievement of children with intellectual disability: The role of socioeconomic status, placement, and parents' engagement. *Res. Dev. Disabil.* **2012**, *33*, 1615–1625. [CrossRef] [PubMed]

52. Stanovich, K.E. Matthew effects in reading: Some consequences of individual differences in the acquisition of literacy. *Read. Res. Q.* **1986**, *21*, 360–407. [CrossRef]

53. Næss, K.A.; Lervåg, A.; Lyster, S.A.; Hulme, C. Longitudinal relationships between language and verbal short-term memory skills in children with Down syndrome. *J. Exp. Child Psychol.* **2015**, *135*, 43–55. [CrossRef] [PubMed]

54. Klinkenberg, J.; Skaar, E. *STAS. Standardisert Test i Avkoding og Staving. [Standardized Test in Decoding and Spelling]*; Pedagogisk-Psykologisk Tjeneste: Hønefoss, Norway, 2003.

55. Lervåg, A.; Bråten, I.; Hulme, C. The cognitive and linguistic foundations of early reading development: A Norwegian latent variable longitudinal study. *Dev. Psychol.* **2009**, *45*, 764–781. [CrossRef]

56. Wechsler, D. *WPPSI-III: Wechsler Preschool and Primary Scale of Intelligence Manual*; The Psychological Corporation: San Antonio, TX, USA, 2002.

57. Dunn, L.; Dunn, L.; Whetton, C.; Burley, J. *The British Picture Vocabulary Scale*; NFER-Nelson: Windsor, UK, 1997.

58. Lyster, S.; Horn, E.; Ryvgold, A. *Vocabulary and Vocabulary Development in Norwegian Child and Adolescents*; Norsk: Oslo, Norway, 2010.

59. Child Language & Learning's Tests. *Tests Developed to a Longitudinal Study*; Department of Special Needs Education, University of Oslo: Oslo, Norway, 2007.

60. Furnes, B.; Samuelsson, S. Preschool cognitive and language skills predicting Kindergarten and Grade 1 reading and spelling: A cross-linguistic comparison. *J. Res. Read.* **2009**, *32*, 275–292. [CrossRef]

61. Gathercole, S.E.; Baddeley, A.D. *The Children's Test of Nonword Repetition*; Psycological Corporation: London, UK, 1996.

62. Wechsler, D. *Wechsler Preschool and Primary Scale of Intelligence—Revised*; Psychological Corporation: San Antonio, TX, USA, 1989.

63. Newton, M.; Thomson, M. *The Aston Index: A Screening Procedure for Written Language Difficulties*; Learning Development Aids: Wisbech, UK, 1976.

64. Carroll, J.M.; Snowling, M.J.; Stevenson, J.; Hulme, C. The development of phonological awareness in preschool children. *Dev. Psychol.* **2003**, *39*, 913–923. [CrossRef]

65. Chen, H.; Cohen, P.; Chen, S. How big is a big odds ratio? Interpreting the magnitudes of odds ratios in epidemiological studies. *Commun. Stat. Simul. Comput.* **2010**, *39*, 860–864. [CrossRef]

66. Ministry of Education. Rammeplan for Barnehagens Innhold og Oppgaver. Available online: https://www.udir.no/laring-og-trivsel/rammeplan/ (accessed on 1 March 2021).

67. Steele, A.; Scerif, G.; Cornish, K.; Karmiloff-Smith, A. Learning to read in Williams syndrome and Down syndrome: Syndrome-specific precursors and developmental trajectories. *J. Child Psychol. Psychiatry* **2013**, *54*, 754–762. [CrossRef]

68. van Bysterveldt, A.; Gillon, G. A descriptive study examining phonological awareness and literacy development in children with Down syndrome. *Folia Phoniatr. Logop.* **2014**, *66*, 48–57. [CrossRef]

69. Muter, V.; Hulme, C.; Snowling, M.J.; Stevenson, J. Phonemes, rimes, vocabulary, and grammatical skills as foundations of early reading development: Evidence from a longitudinal study. *Dev. Psychol.* **2004**, *40*, 665–681. [CrossRef]

70. Nation, K. Children's reading difficulties, language, and reflections on the simple view of reading. *Aust. J. Learn. Difficulties* **2019**, *24*, 47–73. [CrossRef]
71. Foulin, J.N. Why is letter-name knowledge such a good predictor of learning to read? *Read. Writ.* **2005**, *18*, 129–155. [CrossRef]
72. Torgesen, J.K.; Wagner, R.K.; Rashotte, C.A. Longitudinal studies of phonological processing and reading. *J. Learn. Disabil.* **1994**, *27*, 276–286. [CrossRef] [PubMed]
73. Gustavson, K.; von Soest, T.; Karevold, E.; Røysamb, E. Attrition and generalizability in longitudinal studies: Findings from a 15-year population-based study and a Monte Carlo simulation study. *BMC Public Health* **2012**, *12*, 918. [CrossRef] [PubMed]

Article

Symptoms of Autism Spectrum Disorder in Individuals with Down Syndrome

Amanda Dimachkie Nunnally [1,*], Vivian Nguyen [1], Claudine Anglo [1], Audra Sterling [2], Jamie Edgin [3], Stephanie Sherman [4], Elizabeth Berry-Kravis [5], Laura del Hoyo Soriano [1], Leonard Abbeduto [1,6] and Angela John Thurman [1,6]

[1] MIND Institute, University of California Davis Health, 2825 50th Street, Sacramento, CA 95816, USA; vivng@ucdavis.edu (V.N.); Claudine.Anglo@ucsf.edu (C.A.); ldelhoyo@ucdavis.edu (L.d.H.S.); ljabbeduto@ucdavis.edu (L.A.); ajthurman@ucdavis.edu (A.J.T.)
[2] Waisman Center and Department of Communication Sciences and Disorders, University of Wisconsin-Madison, Madison, WI 53706, USA; audra.sterling@wisc.edu
[3] Department of Psychology, University of Arizona, Tucson, AZ 85721, USA; jedgin@email.arizona.edu
[4] Department of Human Genetics, School of Medicine, Emory University, Atlanta, GA 30322, USA; ssherma@emory.edu
[5] Departments of Pediatrics, Neurological Sciences and Biochemistry, Rush University Medical Center, Chicago, IL 60612, USA; elizabeth_berry-kravis@rush.edu
[6] Department of Psychiatry and Behavioral Sciences, University of California, Davis Health, Sacramento, CA 95817, USA
* Correspondence: adimachkie@ucdavis.edu

Citation: Dimachkie Nunnally, A.; Nguyen, V.; Anglo, C.; Sterling, A.; Edgin, J.; Sherman, S.; Berry-Kravis, E.; del Hoyo Soriano, L.; Abbeduto, L.; Thurman, A.J. Symptoms of Autism Spectrum Disorder in Individuals with Down Syndrome. *Brain Sci.* **2021**, *11*, 1278. https://doi.org/10.3390/brainsci11101278

Academic Editor: Margaret B. Pulsifer

Received: 15 July 2021
Accepted: 23 September 2021
Published: 26 September 2021

Publisher's Note: MDPI stays neutral with regard to jurisdictional claims in published maps and institutional affiliations.

Abstract: There is a growing body of evidence to suggest that individuals with Down syndrome (DS) are diagnosed with autism spectrum disorders (ASD) at a higher rate than individuals in the general population. Nonetheless, little is known regarding the unique presentation of ASD symptoms in DS. The current study aims to explore the prevalence and profiles of ASD symptoms in a sample of individuals with DS (n = 83), aged between 6 and 23 years. Analysis of this sample (M_{Age} = 15.13) revealed that approximately 37% of the sample met the classification cut-off for ASD using the Autism Diagnostic Observation Schedule 2 (ADOS-2) Calibrated Severity Score (CSS), an indicator of the participants' severity of ASD-related symptoms. Item-level analyses revealed that multiple items on Module 2 and Module 3 of the ADOS-2, mostly in the Social Affect (SA) subdomain, differentiated the children with DS who did not meet ASD classification (DS-only) from those who did (DS + ASD). Lastly, comparisons of individuals with DS-only and those with DS + ASD differed significantly on the syntactic complexity of their expressive language. These findings shed light on the unique presentation of ASD symptoms in a sample of individuals with DS and suggest that expressive language abilities may play a pivotal role in the presentation of ASD symptoms in DS.

Keywords: Down syndrome; autism spectrum disorder; co-occurring; prevalence

1. Introduction

Down syndrome (DS) is caused by the presence of a third copy of all or part of chromosome 21 and is the leading genetic cause of intellectual disability (ID), affecting approximately 1 in 700 individuals born in the United States [1]. Individuals with DS have historically been described as particularly affable and sociable [2], leading to the belief that they do not experience substantial challenges in the social domain. This belief, however, has been challenged by findings of delays in the development of social communication and social cognition associated with DS, detected as early as infancy [3–9]. In addition to social communication delays in this population, researchers have also reported higher rates of restricted and repetitive interests and behaviors [5,6]. Although the combination of challenges in social communication and rigid and repetitive interests and behaviors is most often associated with autism spectrum disorders (ASD), there is also research to suggest that these symptoms also present among individuals with DS at low risk for ASD [5,6], likely as a reflection of the cognitive and linguistic delays associated with the DS phenotype [10–12]. Studies are needed to clarify whether the social affective challenges and restricted and repetitive interests and behaviors in individuals with DS are best viewed as symptoms of ASD or of the DS phenotype more generally.

Although variable findings regarding the prevalence of ASD among individuals with DS have been reported (16–42%) [13–15], nearly all studies have reported a prevalence higher than the 1.9% (i.e., 1 in 54) prevalence rate observed in the general population [16]. The studies reporting on the prevalence of ASD symptomatology among individuals with DS have differed in the instruments used to ascertain symptoms in this population, with studies utilizing direct assessment methods (e.g., Autism Diagnostic Observation Schedule-2 (ADOS-2) [17]) and/or parent report measures (e.g., Autism Diagnostic Interview-Revised (ADI-R) [17–19], Social Communication Questionnaire (SCQ) [3,15,20], Social Responsiveness Scale (SRS) [5,6], and Aberrant Behavior Checklist (ABC) [21,22]). These data provide a starting point for understanding the nature of social affective skills and restricted and repetitive interests and behaviors among individuals with DS.

There is a relatively small body of work that has demonstrated differences in the presentation of social communication challenges and rigid and repetitive interests and behaviors when comparing individuals with DS + ASD with individuals with DS without co-occurring ASD (referred to hereafter as DS-only). More specifically, researchers using the Aberrant Behavior Checklist (ABC) have found that individuals with DS + ASD present with higher levels of stereotypy and repetitive behaviors than their counterparts with DS-only [21,22]. Similar results were presented by researchers using the ADI-R to compare ASD symptomatology in individuals with DS and those with DS + ASD, finding that individuals with DS + ASD had elevated scores on the reciprocal social interaction, communication and restricted, repetitive, and stereotyped patterns of behavior subdomains when compared to their peers with DS matched on mental-age (MA) [18,19]. These findings have also been replicated using direct observation measures, including in the only study to use the ADOS-2, finding more rigid and repetitive behaviors and greater social communication challenges among individuals with DS + ASD in comparison to those with DS-only [17]. Collectively, these findings suggest more severe ASD symptomatology among individuals with DS + ASD in comparison to those with DS-only. Additionally, although differences have been found between the ASD symptomatology exhibited by individuals with DS + ASD and those with DS-only, it is important to note that studies have found that individuals with DS-only still present with elevated rates of ASD symptomatology when comparing their scores to normative sample means [5,6].

In addition to differences in ASD symptomatology among individuals with DS-only and DS + ASD, group differences have also been found in terms of other dimension of functioning. First, individuals with DS + ASD have been found to have lower cognitive abilities when compared to those with DS-only [5,17,19]. Additionally, differences in expressive and receptive language abilities have been reported, such that individuals with DS + ASD have lower expressive and receptive language abilities than those with DS-only [17,19]. These findings further bolster the notion that individual differences in important domains of ability, such as language, may play a role in the presentation of ASD symptomatology among individuals with DS + ASD.

Altogether, this body of research begins to provide an understanding of the overall nature of social affective skills and restricted and repetitive behaviors and interests among individuals with DS; however, further research is needed to better understand the specific profile of ASD symptomatology in this population. Improving understanding of the presentation of ASD in individuals with DS could lead to earlier and more accurate ASD classification in this population, allowing for earlier access to intervention services which could improve long-term outcomes. Furthermore, the ability to discriminate between symptoms and behaviors that are phenotypic of DS and those which are indicative of ASD would be helpful in determining the type of intervention needed to facilitate improved outcomes.

The purpose of the current study was to use a direct assessment, gold-standard autism diagnostic instrument (i.e., the ADOS-2) in a sample of individuals with DS to: (1) explore the proportion of individuals who meet criteria for ASD diagnosis, (2) determine whether individuals who do and do not meet criteria for ASD diagnosis differ on key individual characteristics such as cognitive and linguistic ability, and (3) investigate whether specific items on Module 2 and Module 3 of the ADOS-2, which are designed to be administered to individuals at different developmental levels, differentiated those who met criteria for ASD diagnosis from those who did not.

2. Materials and Methods

Study procedures were reviewed and approved by Institutional Review Boards at all participating universities. Written informed consent was obtained from participants' guardians, and verbal assent was obtained from the youth prior to beginning study procedures. All data for the present study were collected at participants' initial visit.

2.1. Participants

The sample for the current study was drawn from a larger sample of individuals with DS, aged between 6 and 23 years, who were recruited as part of a multi-site study evaluating the feasibility of expressive language sampling (ELS) as an outcome measure [23,24]. The chronological age range of the larger ELS project was selected to include individuals who would likely be able to meaningfully complete the ELS tasks and exclude those who might display clinically significant signs of Alzheimer's Disease. All participants provided medical documentation of Down syndrome (i.e., trisomy 21 or translocation) without mosaicism, and all met criteria for ID. In addition, the following inclusion criteria were utilized in the larger study, based on parent report: (1) participant and caregiver willingness to partake in the protocol; (2) participants' use of speech as their primary mode of communication, with the use of at least occasional multi-word utterances; (3) participants' use of English as their primary language; (4) no more than mild hearing loss; (5) no serious (uncorrected) visual impairment that may interfere with participants' performance on the testing battery; (6) participants' IQs fell within the range for ID ($\leq$70) and (7) participants were not enrolled in a randomized control trial or experiencing medication, treatment or significant educational changes during the 8 weeks prior to the initial testing visit.

In the larger study, participants were recruited and tested at four university sites, located in Arizona, Georgia, California and Wisconsin, although many participants resided outside these states. The total sample of 107 participants with DS (55 males, 52 females; M_{Age} = 15.13). Participants in the present study were excluded from or analysis if they had missing or incomplete ADOS-2 (n = 13) or if they received Module 1 of the ADOS-2 (n = 6). The decision to exclude participants who received Module 1 of the ADOS-2 from analyses was based on the study inclusion criteria, which was meant to target participants with at least multi-word utterances. Because these eligibility criteria would have excluded most participants with DS for whom Module 1 of the ADOS-2 was chosen, as they would not meet the threshold for language, these data were not considered representative of the larger population. The final sample for the present study was thus comprised of 83 participants (45 males, 38 females) with a mean age of 15.54 years (SD = 5.19) and a mean Stanford-Binet Intelligence Scales, Fifth Edition (SB-5) Full Scale IQ deviation score of 46.65 (SD = 11.21). The choice to use deviation scores in this study was due to the large number of participants that received the lowest possible score (floor scores) on FSIQ. Deviation scores provide z-score transformations based on population norms and are useful in ameliorating floor effects and allow for a more accurate measure of cognitive abilities among individuals with ID [25].

Because some of the research objectives consider participant performance as a function of ADOS-2 module, participant characteristics were also considered as a function of module. Participants who received Module 2 were, on average, younger than those who received Module 3 ($F(1,81) = 16.339$, $p < 0.001$). Additionally, lower VIQ deviation scores ($F(1,73) = 37.748$, $p < 0.001$) and NVIQ deviation scores ($F(1,76) = 22.663$, $p < 0.001$) were observed for participants receiving Module 2 in comparison to those receiving Module 3. Conversely, participants who received Module 2 and those who received Module 3 were not significantly different in their adaptive functioning skills ($F(1,64) = 656$, $p = 0.421$). Please see Table 1 for additional details.

Table 1. Participant demographics for overall sample, Module 2 and Module 3.

	Overall Sample			Module 2			Module 3		
	N	Frequency	%	N	Frequency	%	N	Frequency	%
Gender (M)	83	45	54.2	45	28	62.2	38	17	44.7
Race	83			45			38		
African American/Black		2	2.4		2	4.4		0	0
Asian/Pacific Islander		1	1.2		1	2.2		0	0
White		58	69.9		31	68.9		27	71.1
Multiple Races		9	10.8		6	13.3		3	7.9
Unknown		12	14.5		5	11.1		7	18.4
Other		1	1.2		0	0		1	2.6
Ethnicity	83			45			38		
Hispanic/Latino		16	19.3		8	9.6		8	9.6
Yearly Income									
Less than 25,000		5	6.0		3	7.0		2	6.2
25,000–50,000		18	21.7		13	30.2		5	22.2
50,000–75,000	81	12	14.5	43	8	18.6	38	4	14.8
75,000–100,000		13	15.7		5	11.6		8	16.0
100,000–150,000		15	18.1		6	14.0		9	18.5
150,000–250,000		12	14.5		7	16.3		5	14.8
Over 250,000		6	7.2		1	2.3		5	7.4
	N	Mean	SD	N	Mean	SD	N	Mean	SD
CA (years)	83	15.60	5.18	45	13.66	5.35	38	17.89	3.93
Cognitive									
FSIQ Deviation	71	46.66	11.21	38	40.77	8.53	33	53.44	10.12
NVIQ Deviation	77	50.91	11.29	41	45.85	9.81	36	56.68	10.12
VIQ Deviation	74	41.88	12.60	41	35.30	9.67	33	50.06	10.98
ADOS-2									
CSS	83	3.27	2.16	45	3.73	2.19	38	2.71	2.04
SA Severity	83	3.83	2.12	45	4.27	2.05	38	3.32	2.11
RRB Severity	83	2.72	4.27	45	4.31	2.24	38	4.21	2.72
Adaptive Functioning	65			40			25		
Vineland ABC SS		73.69	28.03		75.93	35.00		70.12	9.09

Note: SES = Socioeconomic Status; CA = Chronological Age; FSIQ Deviation = Full Scale IQ Deviation; NVIQ Deviation = Non-Verbal IQ Deviation; VIQ Deviation = Verbal IQ Deviation; CSS = Calibrated Severity Score; SA Severity = Social Affective Severity Score; RRB Severity = Rigid and Repetitive Behavior Severity Score; Vineland ABC SS = Adaptive Behavior Composite Standard Score.

2.2. Measures

2.2.1. Cognitive Ability

Participants' cognitive ability was assessed using the Stanford Binet Intelligence Scales, Fifth Edition (SB-5) [26]. Deviation scores were calculated to provide descriptive information on the study sample for Full Scale IQ (FSIQ), Non-Verbal IQ (NVIQ), Verbal IQ (VIQ), following procedures outlined by Sansone and colleagues [26]. Deviation scores, which provide z-score transformation based on the general population norms, are helpful in mitigating floor effects and lend themselves to a more precise measurement of cognitive ability in populations with ID [25]. In addition, Non-Verbal Change Sensitive scores, the equivalent of growth scores, were calculated for use in study analyses.

2.2.2. ASD Symptom Severity

The Autism Diagnostic Observation Schedule-2 is a semi-structured, standardized play-based assessment used to measure reciprocal interactions and repetitive behaviors. Participants in this sample received either Module 2 ($n = 45$) or Module 3 ($n = 38$) of the ADOS-2, administered by examiners trained to research reliability. In addition, site examiners scored video administrations and participated in cross-site pre-collection reliability calls to calibrate scoring and cross-site reliability was also assessed for 13 DS administrations collected on the project. Administrator reliability was 86% for all items and 87% for algorithm items. ADOS-2 modules were assigned based on participants' verbal ability following the published guidelines for the measure, such that participants with "phrase speech up to fluent speech" received Module 2, and those who are "producing a range of flexible sentence types, providing language beyond the immediate context, and describing logical connections within a sentence" [27] (p. 10) received Module 3. Due to the level of developmental delay exhibited by participants, no participants demonstrated "a minor level of independence in relationships and goals" [27] (p. 11) required to receive Module 4 of the ADOS-2. For the purposes of this study, the overall calibrated severity score (Overall CSS), Social Affect calibrated severity score, (SA-CSS), and Restricted and Repetitive Behavior calibrated severity score (RRB-CSS) were calculated to provide standardized scores for symptom severity [28]. Both the Overall CSS and SA-CSS are assessed using a 10-point scale. In contrast, the RRB-CSS score is assessed using a 7-point scale that was spread across a 10-point scale range, in which the scores 2, 3, and 4 are not possible to obtain [29]. Participants' ASD classification was determined using Overall CSS, in accordance with procedures outlined by Gotham, Pickles and Lord [28]. Finally, for participants who were older than the norming sample of the ADOS-2, the upper age limit of the CSS norming tables was used to compute CSSs.

2.2.3. Expressive Language Sampling

Participants' expressive syntactic and lexical levels were assessed using a narration task in which participants were asked to narrate a story using a wordless picture book [23,24]. The task begins with the participant familiarizing themselves with the book by examining each page spread for approximately 10 s before narrating the story depicted in the book. The examiner facilitates the narration by controlling the book and waiting for the participant to finish their description before turning the page. In order to standardize the task, participants received one of two books from the Mercer Mayer's "Frog" series ("Frog Goes to Dinner" or "Frog on His Own"), and administrators relied on a standardized set of prompts and responses to ensure minimal and consistent scaffolding across participants. Participants' speech was transcribed, segmented into C-units (communication units), with a C-unit is defined as an independent clause with associated modifiers, including dependent clauses, and analyzed using the software program Systematic Analysis of Language Transcripts 18 Research (SALT) [30]. Inter-transcriber agreement data were computed and averaged as follows: 87% for utterance segmentation, 87% for identification of partly or fully unintelligible C-units, 84% for identification of the exact lexical and morphemic content of each C-unit, 76% for identification of the exact number of morphemes in each C-unit and 80% for the exact number of words in each C-unit [24]. Construct validity has been established for the ELS narration task such that medium to strong convergent validity was found with directly administered and informant report measures for similar constructs measuring syntactic complexity and lexical diversity in individuals with DS as well as other forms of ID [23,24].

Syntactic complexity. Participants' syntactic maturity was assessed by calculating the mean number of morphemes per C-unit. Only complete and fully intelligible C-units were used to calculate this variable.

Lexical Diversity. The size of each participant's expressive vocabulary was computed by calculating the number of different word roots in the participant's first 50 complete and fully intelligible C-units. In the event that the participant produced less than 50 complete and fully intelligible C-units, the full sample was used.

2.3. Data Analysis

To address Objective 1, the frequency of ASD classification (i.e., the number of individuals who had an overall CSS that met the cutoff for ASD) is presented for the overall sample, as well by module. One-way ANOVAs were conducted to compare the frequency of classification in Module 2 and Module 3. Additionally, distributions for CSS on the SA and RRB subdomains were also presented for the overall sample, and one-way ANOVAS were conducted to determine whether group differences (DS + ASD vs. DS-only) between SA-CSS and RRB-CSS were detected in the overall sample. Next, to address Objective 2, analyses compared key individual characteristics (chronological age, nonverbal change sensitive score, syntactic complexity and lexical diversity) between participants with DS who did not receive an ASD classification on the ADOS-2 (DS-only) and participants with DS who received an ASD classification on the ADOS-2 (DS + ASD). Parametric analyses (one-way ANOVAs) were used to address Objective 3 due to the continuous nature and normal distribution of participant characteristic. Lastly, to address Objective 3, we explored group differences between individuals with DS-only and DS + ASD across ADOS-2 algorithm items, doing so separately for each of the ADOS-2 modules since items differ between Module 2 and Module 3 of the ADOS-2. Because of the ordinal nature and non-normal distribution of ADOS-2 algorithm item scores, nonparametric analyses (Mann–Whitney U-Tests) were used. False Discovery Rate (FDR) corrections were applied within each set of analyses, in accordance with procedures outlined by Benjamini and Hochberg [31], to maintain a family-wise alpha rate of $p \leq 0.050$.

3. Results

3.1. Prevalence of ASD

ASD Classification

In the current sample, 37.3% of participants met the overall classification criteria for ASD on the ADOS-2 (i.e., DS + ASD). The prevalence of ASD was higher among those receiving Module 2 (46.7%) than among those receiving Module 3 (26.3%); statistical comparisons indicated that this difference in rate between modules approached significance ($F(1,82) = 3.722$; $p = 0.057$; $\eta^2 = 0.044$). See Figure 1 for distribution of participant scores.

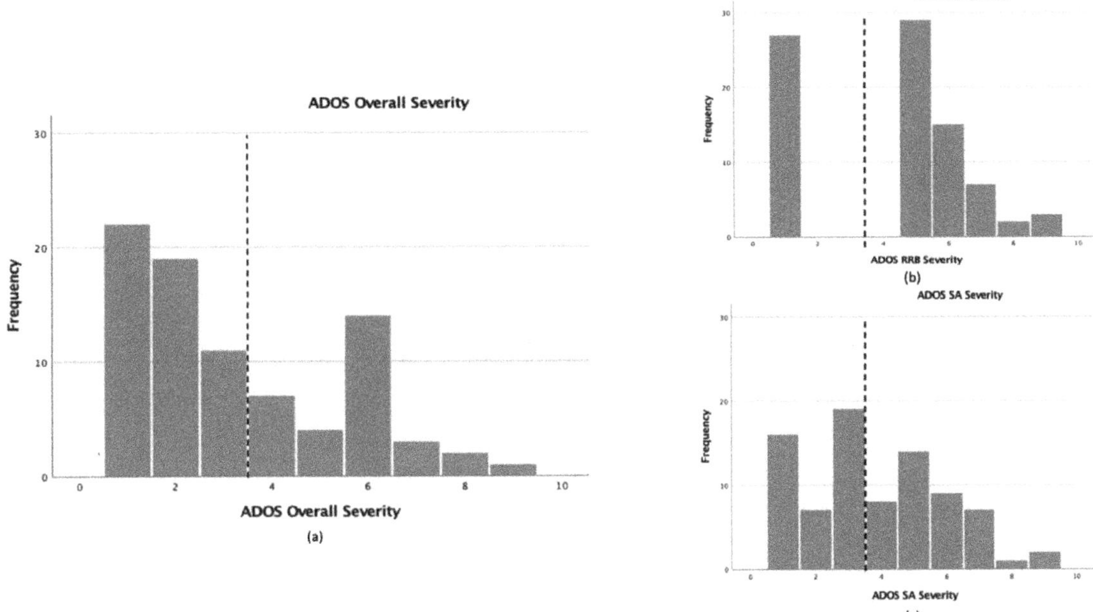

Figure 1. Frequency distribution of ADOS-2 CSSs across sample; (**a**) ADOS-2 Overall CSS for overall sample; (**b**) ADOS-2 RRB-CSS for overall sample; (**c**) ADOS-2 SA-CSS for overall sample. Note: Dotted line represents CSS cutoff for ASD classification.

Analyses were also conducted to determine whether participants classified as DS + ASD differed from participants classified as DS-only on CSS scores for the SA and RRB subdomains. In the overall sample, group differences were detected in both the SA ($F(1,82) = 151.740$; $p < 0.001$; $\eta^2 = 0.652$) and RRB domains ($F(1,82) = 20.115$; $p < 0.001$; $\eta^2 = 0.199$), such that individuals with DS + ASD had higher scores than those with DS-only. When exploring group differences at the module level, significant differences were detected for both SA-CSS ($F(1,44) = 76.495$; $p < 0.001$; $\eta^2 = 0.640$) and RRB-CSS for Module 2 ($F(1,44) = 35.350$; $p < 0.001$; $\eta^2 = 0.451$); however, only SA-CSS significantly differentiated groups in Module 3 ($F(1,37) = 64.184$; $p < 0.001$; $\eta^2 = 0.641$). Please see Figure 2 for means.

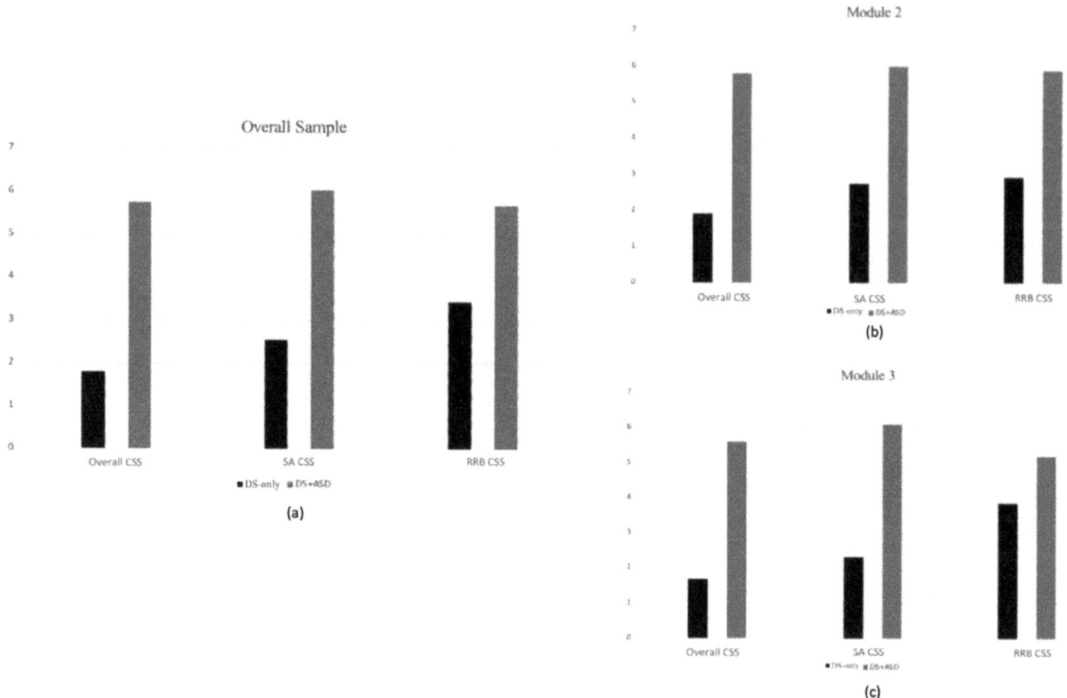

Figure 2. Group differences in mean ADOS-2 CSSs; (**a**) Group differences in mean ADOS-2 CSSs for overall sample; (**b**) Group differences in mean ADOS-2 CSSs for Module 2; (**c**) Group differences in mean ADOS-2 CSSs for Module 3.

3.2. Group Differences across Characteristics

We compared the participants classified as DS + ASD to participants classified as DS-only in terms of chronological age, nonverbal cognitive ability, lexical diversity, and syntactic complexity (see Table 2). Significant group differences were found for nonverbal cognitive ability ($F(1,82) = 1.091$; $p = 0.044$; $\eta^2 = 0.053$), lexical diversity ($F(1,82) = 7.330$ $p = 0.008$; $\eta^2 = 0.085$), and syntactic complexity ($F(1,82) = 4.198$; $p = 0.000$; $\eta^2 = 0.168$), such that group means were lower for participants classified as DS + ASD than for participants classified as DS-only across all comparisons. The differences in lexical diversity and syntactic complexity, but not nonverbal cognitive ability, remained significant after applying the FDR correction.

Follow up analyses were conducted to determine whether differences found in lexical diversity and syntactic complexity were also seen for each of the modules analyzed separately. Results of these analyses revealed that, within Module 2, participants classified as DS + ASD produced C-units with less syntactic complexity ($F(1,42) = 7.095$ $p = 0.011$; $\eta^2 = 0.148$) than participants classified as DS-only. Similar results were found for Module 3, with a lower mean for syntactic complexity ($F(1,37) = 5.201$ $p = 0.029$; $\eta^2 = 0.126$) found for participants classified as DS + ASD than DS-only. No significant differences in lexical diversity was found between individuals who were classified as DS + ASD and those classified as DS-only in either Module 2 ($F(1,42) = 2.484$ $p = 0.123$; $\eta^2 = 0.057$) or Module 3 ($F(1,37) = 1.125$ $p = 0.296$; $\eta^2 = 0.030$).

Table 2. Means for Participants Characteristics for overall sample, Module 2 and Module 3.

		Overall Sample			Module 2			Module 3		
		N	Mean	SD	N	Mean	SD	N	Mean	SD
CA	DS-only	52	15.14	5.21	24	11.87	4.77	28	17.93	3.78
	DS + ASD	31	16.36	5.14	21	15.69	5.36	10	17.78	4.57
	Total	83	15.60	5.19	45	13.66	5.35	38	17.89	3.93
SB-5 NV Change Sensitive Score	DS-only	47	464.47	13.80	21	455.95	11.63	26	471.35	11.50
	DS + ASD	30	458.07	12.65	20	454.95	12.95	10	464.30	9.87
	Total	77	461.97	13.65	41	455.46	12.15	36	469.39	11.39
Syntactic Complexity	DS-only	51	5.18	2.02	23	3.76	1.61	28	6.35	1.533
	DS + ASD	30	3.38	1.86	20	2.52	1.42	10	5.09	1.40
	Total	81	4.51	2.14	43	3.19	1.63	38	6.02	1.58
Lexical Diversity	DS-only	51	72.71	37.09	23	49.17	32.26	28	92.04	28.98
	DS + ASD	30	50.63	32.38	20	35.50	23.07	10	80.90	26.99
	Total	81	64.53	36.81	43	42.81	28.87	38	89.11	18.54

Note: CA = Chronological age; SB-5 NV Change Sensitive Score = Stanford Binet-5 Non-Verbal Change Sensitive Score.

3.3. Group Differences across ADOS-2 Items

3.3.1. Module 2 Items

Between-group comparisons of algorithm items on Module 2 of the ADOS-2 revealed 7 items in the SA domain and 3 items in the RRB domain that differentiated participants with DS + ASD from participants with DS-only (see Figure 3). The differences on all of these items remained significant after the FDR correction. More specifically, differences in group mean ranks were detected for SA algorithm items measuring: (1) descriptive, conventional, instrumental or informational gestures (U = 160.50; p = 0.012); (2) unusual eye contact (U = 108.00; p < 0.001); (3) facial expressions directed to others (U = 122.00; p = 0.001); (4) showing (U = 74.50; p < 0.001); (5) quality of social overtures (U = 133.50; p = 0.002) ; (6) amount of reciprocal social communication (U = 157.50; p = 0.017); (7) overall quality of rapport (U = 92.50; p < 0.001). Additionally, differences in group mean ranks were detected for RRB algorithm items measuring: (1) stereotyped/Idiosyncratic use of words or phrases (U = 172.50; p = 0.029); (2) hand and finger and other complex mannerisms (U = 188.50; p = 0.002); (3) unusually repetitive interests or stereotyped behaviors (U = 150.50; p = 0.007).

3.3.2. Module 3 Items

Between-group comparisons were also conducted for algorithm items on Module 3 of the ADOS-2 (see Figure 4). Group differences were found in the mean ranks for the five SA algorithm items measuring: (1) reporting of events (U = 75.500; p = 0.016); (2) unusual eye contact (U = 76.00; p = 0.005); (3) quality of social overtures (U = 54.50; p = 0.001); (4) quality of social response (U = 76.00; p = 0.017); (5) amount of reciprocal social communication (U = 38.00; p < 0.001); (6) overall quality of rapport (U = 43.00; p < 0.001). These differences remained significant even after applying the FDR corrections. One algorithm item from the RRB subdomain, measuring unusually repetitive interests or stereotyped behaviors, emerged as significantly different between participants who met classification from those who did not (U = 86.50; p = 0.034); however, this finding did not remain significant after applying the FDR correction.

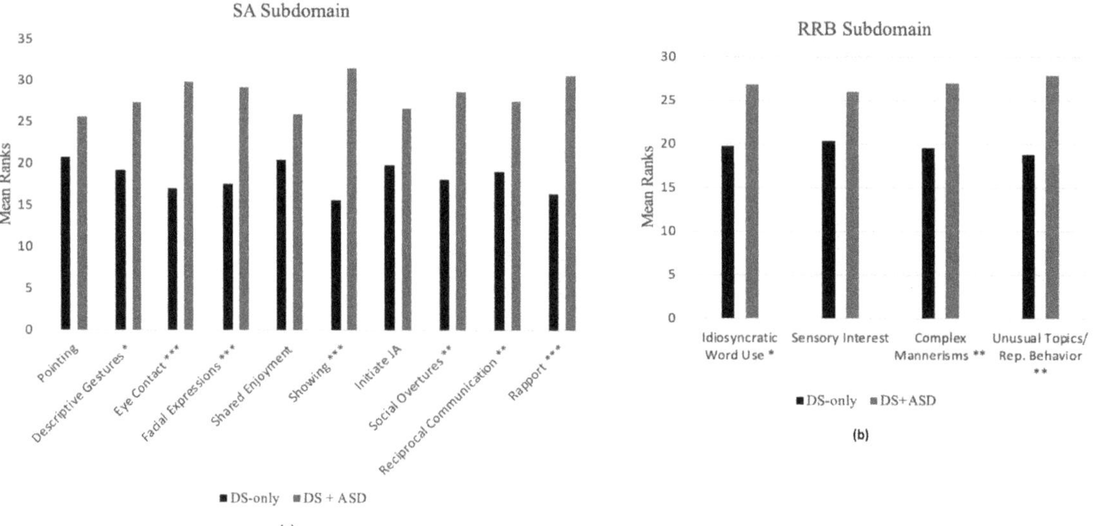

Figure 3. Mean Rank comparisons for Module 2 algorithm items; (**a**) Mean Rank comparisons for Module 2 SA items; (**b**) Mean Rank comparisons for Module 2 RRB items. Note: JA = Joint attention. * $p < 0.05$; ** $p < 0.01$; *** $p < 0.001$.

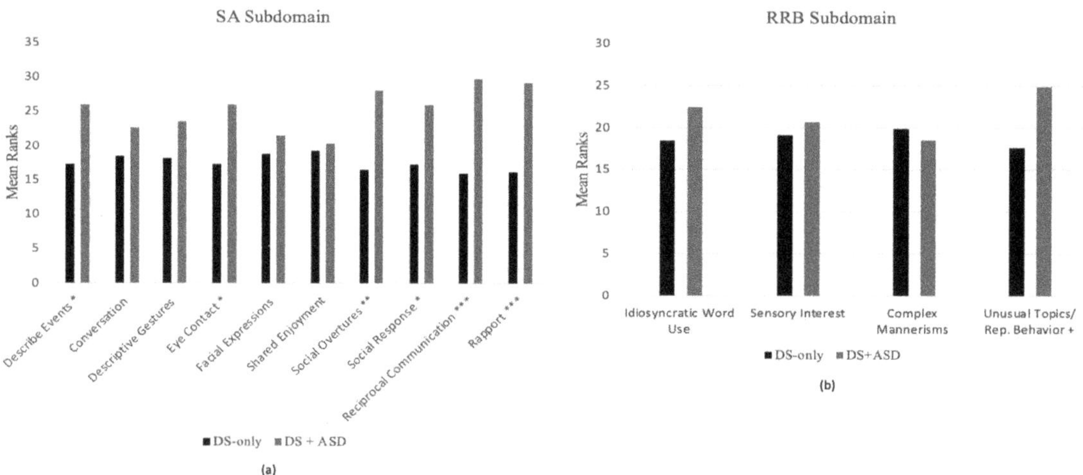

Figure 4. Mean Rank comparisons for Module 3 algorithm items; (**a**) Mean Rank comparisons for Module 3 SA items; (**b**) Mean Rank comparisons for Module 3 RRB items. Note: * $p < 0.05$; ** $p < 0.01$; *** $p < 0.001$; + signifies no longer significant after FDR correction.

4. Discussion

The detection of co-occurring ASD among individuals with ID associated with known genetic conditions, such as DS, poses many challenges [32,33]. It is becoming increasingly apparent that individuals with DS are at increased risk for presenting with the symptoms ASD relative to the general population [5,6]; moreover, due to the developmental delays associated with the DS phenotype may influence the presentation of ASD symptomatology in this population [10–12]. Studies that clarify the nature of social communication skills and restricted and repetitive interests and behaviors in DS, and other populations with ID, can provide important information for understanding the unique way in which ASD presents among individuals. The goal of the present study was to elucidate the prevalence and the factors shaping the presentation of ASD symptomatology in a large sample of individuals with DS using the ADOS-2, a gold standard direct-assessment diagnostic instrument.

4.1. ASD Classification

Several key findings emerged from the present study. First, we explored the prevalence of ASD in a sample of 83 individuals with DS and found that 37.3% of the sample met overall classification criteria for ASD on the ADOS-2, which falls within the range of prevalence rates presented in several previous publications on DS [13–15]. This similarity in prevalence rates suggests that the specific measures used may not differ significantly in their utility in detecting ASD in individuals with DS. Individuals in the sample who had overall scores that met classification cutoff for ASD (DS + ASD) had significantly higher SA-CSS and RRB-CSS scores than individuals who did not (DS-only). This finding is consistent with prior research finding of both more challenges in social communication and increased rigidity and repetitive behaviors when comparing individuals classified as having DS-only and those classified as having DS + ASD [18–22]. It should be noted that studies have found that individuals with DS who were at low risk for ASD nonetheless presented with challenges in social communication and restricted and repetitive interests and behaviors relative to normative expectations for their chronological ages, indicating that these symptoms and behaviors may also be phenotypic to DS [5,6]. This is further underscored by findings in the current study that a number of individuals received overall scores on the ADOS-2 that were right below the cutoff for ASD classification.

Closer examination of the data at the module level indicated a trend for overall ASD classification rates to be almost two times higher among those individuals receiving the ADOS-2 Module 2 than on those receiving Module 3. Group differences were detected in Module 2, with individuals with DS + ASD having significantly higher means on both the SA-CSS and RRB-CSS than those with DS-only. There are two possible explanations for the differences seen in ASD classification rates and extent of the differences between individuals receiving the two modules. First, the finding of overall lower cognitive and linguistic abilities amongst individuals who received Module 2, which has been shown to be related to ASD symptomatology in this population could be driving this effect [5,17,18]. Second, it is possible that particular items, activities, and/or norming procedures associated with the Module 2 are contributing to these differences [34–36]. In other words, the items specific to Module 2 may be more sensitive to the comparisons made in the present study, therefore, considering item analyses and the influence of participant characteristics on classification rates provide a start to clarifying these findings. In addition to differences in classification rates between modules, differences emerged such that for individuals who received Module 3, only the SA-CSS differentiated individuals with DS + ASD from those with DS-only, with the former having significantly higher scores. There may be several explanations for the RRB-CSS not differentiating groups in Module 3. First, since fewer individuals receiving Module 3 had an RRB-CSS that met the cutoff for ASD classification, between-group comparisons within Module 3 could be underpowered. Second, stereotyped behaviors, highly restricted interests, rigidity, and inflexibility are common in DS [5,18] and may be best viewed as an inherent part of the DS phenotype rather than being reflective of ASD. Given the overlap between seemingly phenotypic rigid and repetitive behaviors seen among individuals with DS and those associated with the core deficits of ASD, further research is necessary to better understand whether these behaviors can be considered indicative of co-occurring ASD in this population or whether they instead reflect different underlying mechanisms and challenges.

4.2. Group Differences across Individual Characteristics

We explored whether differences in chronological age, nonverbal cognitive abilities, and language skills differentiated individuals who were classified as DS + ASD from those classified as DS-only. It is important to note that the group classification was based on

ADOS-2 cutoff scores and does not imply a formal (clinical) diagnosis of ASD. Instead, these groups simply reflect individuals' presentation of ASD symptomatology on the ADOS-2. With that said, when examining groups differences in the overall sample, significant differences (after correcting for multiple comparisons) were identified in lexical diversity and syntactic complexity were detected, such that individuals with DS-only performed at a higher level than those with DS + ASD. Results of follow up analyses indicated that while syntactic complexity significantly differentiated individuals with DS + ASD from those with DS-only in both Module 2 and Module 3, lexical diversity was not a significant difference at the individual module level. One possible explanation for the finding that lexical diversity differentiated groups in the overall sample but not at the module level may be that the difference detected may be more appropriately attributed to ID associated with DS than to ASD symptomatology. This is supported by the fact that individuals who received Module 2 had, on average, lower cognitive and linguistic abilities than those who received Module 3. Overall, the finding of lower linguistic abilities among individuals with DS + ASD in the overall sample is consistent with previously reported findings of lower verbal abilities among individuals with DS + ASD than those with DS-only [19,21]. Moreover, these findings underscore the need to consider the linguistic delays characteristic of individuals with DS in the interpretation of ADOS-2 algorithm items, especially with samples of older individuals with ASD, as studies to derive algorithm items were originally normed using samples of individuals aged up to 12 years [34,35].

4.3. Group Differences across ADOS-2 Items

We explored the presentation of ASD symptomatology in this sample by comparing groups (i.e., DS + ASD and DS-only) across algorithm items of the ADOS-2. Analyses were conducted at the module level because algorithm items differ between modules. Items on the SA subdomain of the ADOS-2 were more likely to differentiate individuals with DS + ASD from those with DS-only, with four common items emerging across modules: unusual eye contact, quality of social overtures, amount of reciprocal social communication and overall quality of rapport. Of note, all four significant algorithm items in the SA subdomain rely on pragmatic social communication skills, which have been found to be delayed among young children with DS and other forms of ID relative to normative age expectations. In previous studies, children with DS have also been found to have a relative weakness in pragmatics in comparison with their structural language abilities [37]. Additionally, the finding that unusual eye contact differentiated the groups is of interest given that atypical eye contact in DS has been detected as early as infancy [4] and the use of eye contact among children with DS is less clear than among their typically developing (TD) peers and peer with other developmental disabilities (DD) [38]. These results suggest that challenges in pragmatic social communication are characteristic of the DS phenotype and indicate the need to clarify boundaries between standard phenotype heterogeneity from comorbid ASD symptomatology.

At the same time, many commonalities were observed among SA items that did not differentiate the two DS groups. In Module 2, items measuring pointing, shared enjoyment and joint attention did not significantly differentiate individuals classified as DS-only from those classified as DS + ASD. Similarly, the following four SA items from Module 3 did not significantly differentiate the groups: conversation, descriptive gestures, facial expressions and shared enjoyment. Notably, shared enjoyment did not discriminate between groups in either module, a finding that is not entirely surprising given that, although individuals with DS may struggle with certain aspect of social communication, findings suggest relative strengths in social engagement and social orientation (i.e., sociability) [39]. Additionally, in Module 2, joint attention abilities did not differentiate between participants' group membership. This finding is consistent with literature finding that joint attention may be a relative strength among individuals with DS, who perform similarly to children with TD and at a higher level than children with ASD and other NDDs [40]. Although joint attention (JA) was not explicitly measured in Module 3, it could be argued that joint attention is a developmental antecedent to the use of descriptive gestures, which is measured in Module 3 [41]. The finding that the item measuring the use descriptive gesture also did not differentiate groups may suggest that the use of JA may continue to be a strength among individuals with DS throughout childhood. These results suggest that some phenotypic characteristics of DS, such as relative strengths in social orientation and joint attention, may offset the presentation of these skills among individuals with DS + ASD. Further research is needed to better understand how phenotypic characteristics of DS affect ASD symptom presentation among individuals with DS + ASD.

Three items from the RRB subdomain of the ADOS-2 (stereotyped/idiosyncratic use of words and phrases, hand, finger and other complex mannerisms and unusually repetitive interests or stereotyped behaviors) significantly differentiated individuals with DS-only from those with DS + ASD in Module 2, and (after correcting for multiple comparisons) no item from the RRB subdomain differentiated the groups in Module 3. The finding that scores on the RRB subdomain differentiated groups in Module 2 but not Module 3 is of particular interest, as it mimics earlier reported findings that severity scores in the RRB subdomain did not differentiate individuals who were classified as having DS-only from those classified as having DS + ASD. As explained above, there are several possible explanations for this finding, although further research is needed to better understand this phenomenon.

The findings regarding differences in symptom profile and severity between the DS + ASD and DS-only groups have more general implications for the field. In particular, in examining the high cooccurrence of ASD and fragile X syndrome (FXS) [42], we have argued previously that the categorical diagnosis of ASD can hide mechanistically and clinically important differences among individuals with FXS, and between those with FXS+ASD and those with non-syndromic ASD. Moreover, we have provided empirical support for that claim in several studies providing in-depth analysis of both the ADOS-2 and the ADI-R and in multiple samples of different ages and degree of impairments [43,44]. In the present study, too, we have shown that an ASD diagnosis can "mask" different levels of severity and symptom profiles in individuals with DS as a function of the ADOS-2 module administered. This finding, we believe, has less to do with the specific characteristics of the module administered and is, instead, a reflection of that ways in which ASD symptoms are moderated by the phenotype of DS and within-syndrome variability in that phenotype. Simply focusing on whether an individual with DS meets or does not meet criteria for an ASD diagnosis, therefore, could have the consequences of a failure to understand the factors leading to those symptoms or the best approach to treatment to reduce those symptoms.

4.4. Limitations

Several limitations should be considered when interpreting results of this study. First, prevalence rates reported are based on a sample of convenience rather than using a population-based sample, although it is important to note that the participants were not recruited based on ASD status. With this in mind, caution should be used when interpreting the prevalence rates reported. Second, participants' ASD status was determined solely based on the ADOS-2, making it impossible to determine whether the findings reported here will be replicable using other diagnostic measures. Third, based on inclusion criteria for the original study, which required a specific level of language (i.e., a minimum of at least occasional three-word phrase speech), our findings may not be representative of all individuals with DS, such as those who may be minimally verbal or nonverbal. This emphasizes the need for replication of these findings using samples that include individualizing with a range of language abilities, utilizing the full range of ADOS-2 modules. The inclusion of a sample representing the full range of language abilities would allow a better understanding of how ASD symptomatology present differentially among individuals with differing language abilities. Moreover, our findings of higher rates of classification for youth with more limited language abilities, highlights the importance of considering language skills relative to ASD symptomatology; studies focused on these associations in youth with DS who are in the prelinguistic and first words stages of language development are critical to more fully understand the relation between ASD and language skill. Finally, we did not include other non-DS comparison groups. In particular, it would be useful in future studies of ASD in DS to make comparisons with appropriately matched groups of individuals with non-syndromic ASD and ID of a different origin (e.g., fragile X syndrome). Comparisons between individuals with DS + ASD and matched group of individuals with non-syndromic ASD would allow for a deeper understanding of how the DS phenotypes moderates the expression of ASD and the specific phenotypic factors that play a role in that moderation. Although there is a small body of work that examines the differences in presentation of ASD symptomatology among individuals with DS and those with non-syndromic ASD, e.g., [15,18], further research is needed to truly understand which symptoms are attributable to DS phenotype and which are attributable to ASD among individual with DS + ASD. Similarly, there is a fledgling body of work examining differences between individuals with DS + ASD and matched individuals with ID of other etiologies [45]. Given the dearth of research in this area, future research is needed to are needed help identify which aspects of the expression of ASD that are unique to DS rather than common to those with ID.

5. Conclusions

The findings reported in the current study further elucidate the prevalence of ASD symptomatology in a sample of individuals with DS, as measured by gold standard diagnostic instrument, the ADOS-2. The findings of this study replicate previously reported findings of increased challenges related to social communication and higher levels of rigid and repetitive behaviors among individuals with DS + ASD in comparison to those with DS-only. We also highlight the contribution of language delays to the classification of ASD in this sample, which underscores previously raised questions regarding the boundary between phenotypic characteristics of DS and true ASD symptomatology. Although this study contributes to the field by examining the prevalence and presentation of ASD in the largest sample of individuals with DS to date, it also demonstrates the need for more research exploring the complexities of diagnosing ASD among individuals with DS.

Author Contributions: A.D.N., V.N., C.A., A.S., J.E., S.S., E.B.-K., L.d.H.S., L.A., and A.J.T.; methodology, A.D.N., V.N., C.A., A.S., J.E., S.S., E.B.-K., L.d.H.S., L.A., and A.J.T.; formal analysis, A.D.N., V.N., C.A., L.A., and A.J.T.; investigation, A.D.N., V.N., C.A., A.S., J.E., S.S., E.B.-K., L.d.H.S., L.A., and A.J.T.; data curation, A.D.N., V.N., C.A., A.S., J.E., S.S., E.B.-K., L.d.H.S., L.A., and A.J.T.; writing—original draft preparation, A.D.N., V.N., C.A., L.d.H.S., L.A., and A.J.T.; writing—review and editing, A.D.N., V.N., C.A., A.S., J.E., S.S., E.B.-K., L.d.H.S., L.A., and A.J.T.; visualization, A.D.N.; supervision, A.S., J.E., S.S., E.B.-K., L.d.H.S., L.A., and A.J.T.; project administration, A.D.N., V.N., C.A., A.S., J.E., S.S., E.B.-K., L.d.H.S., L.A., and A.J.T.; funding acquisition, A.S., J.E., S.S., E.B.-K., L.d.H.S., L.A., and A.J.T. All authors have read and agreed to the published version of the manuscript.

Funding: This research was supported by the following grant from the National Institutes of Health, R01HD074346, P50HD103526, and UL1TR001860.

Institutional Review Board Statement: This study was approved by the Institutional Review Board at all participating university sites. The authors assert that all procedures contributing to this work comply with the ethical standards of the relevant national and international committees on human experimentation and with the Helsinki Declaration of 1975, as revised in 2008 (Code: 403210).

Informed Consent Statement: Informed written consent was obtained from the parent/legal guardian prior to participation, and assent was obtained from each participant.

Data Availability Statement: The datasets used and/or analyzed for the present paper can be made available upon a reasonable request to the corresponding author.

Acknowledgments: The authors thank the many staff across all the participating data collection sites who worked on the project. We are indebted to the families for their time, support, and partnership.

Conflicts of Interest: L.A. has received funding from F. Hoffmann-La Roche Ltd., Roche TCRC, Inc., Neuren Pharmaceuticals Ltd., and Lumind to consult on and implement outcome measures in clinical trials for FXS and DS. E.B.-K. has received funding from Seaside Therapeutics, Novartis, Roche, Alcobra, Neuren, Cydan, Fulcrum, GW, Healx, Neurotrope, Marinus, Zynerba, BioMarin, Lumos, Ovid, AMO, Yamo, Ionis, GeneTx, Acadia, Neurogene, Orphazyme, Ultragenyx, Taysha, Tetra, and Vtesse/Sucampo/Mallinkcrodt Pharmaceuticals to consult on trial design or development strategies and/or conduct clinical trials in FXS or other genetic neurodevelopmental or neurodegenerative disorders, and from Asuragen Inc. to develop testing standards for FMR1 testing. A.J.T. has received funding from Fulcrum Therapeutics to develop outcome measures for FXS. The other authors declare that they have no competing interests.

References

1. Mai, C.T.; Isenburg, J.L.; Canfield, M.A.; Meyer, R.E.; Correa, A.; Alverson, C.J.; Lupo, P.J.; Riehle-Colarusso, T.; Cho, S.J.; Aggarwal, D.; et al. National population-based estimates for major birth defects, 2010–2014. *Birth Defects Res.* **2019**, *111*, 1420–1435. [CrossRef] [PubMed]
2. Gibbs, M.V.; Thorpe, J.G. Personality stereotype of noninstitutionalized down syndrome children. *Am. J. Ment. Defic.* **1983**, *87*, 601–605.
3. Channell, M.M.; Project, T.D.S.C.; Hahn, L.; Rosser, T.C.; Hamilton, D.; Frank-Crawford, M.A.; Capone, G.T.; Sherman, S.L. Characteristics Associated with Autism Spectrum Disorder Risk in Individuals with Down Syndrome. *J. Autism Dev. Disord.* **2019**, *49*, 3543–3556. [CrossRef]
4. Hahn, L.J.; Hamrick, L.M.; Kelleher, B.L.; Roberts, J.E. Autism Spectrum Disorder-Associated Behaviour in Infants with Down Syndrome. *J. Health Sci. Educ.* **2020**, *4*, 180.
5. Channell, M.M.; Phillips, B.A.; Loveall, S.J.; Conners, F.A.; Bussanich, P.M.; Klinger, L.G. Patterns of autism spectrum symptomatology in individuals with Down syndrome without comorbid autism spectrum disorder. *J. Neurodev. Disord.* **2015**, *7*, 5. [CrossRef]
6. Channell, M.M. The Social Responsiveness Scale (SRS-2) in school-age children with Down syndrome at low risk for autism spectrum disorder. *Autism Dev. Lang. Impair.* **2020**, *5*. [CrossRef]
7. Fidler, D.J. The Emerging Down Syndrome Behavioral Phenotype in Early Childhood. *Infants Young-Child.* **2005**, *18*, 86–103. [CrossRef]
8. Thurman, A.J.; Mervis, C.B. The regulatory function of social referencing in preschoolers with Down syndrome or Williams syndrome. *J. Neurodev. Disord.* **2013**, *5*, 2. [CrossRef] [PubMed]
9. Kasari, C.; Freeman, S.F.N. Task-Related Social Behavior in Children With Down Syndrome. *Am. J. Ment. Retard.* **2001**, *106*, 253–264. [CrossRef]
10. Chapman, R.S.; Hesketh, L.J. Behavioral phenotype of individuals with Down syndrome. *Ment. Retard. Dev. Disabil. Res. Rev.* **2000**, *6*, 84–95. [CrossRef]
11. Chapman, R.S. Language learning in Down syndrome: The speech and language profile compared to adolescents with cognitive impairment of unknown origin. *Down Syndr. Res. Pract.* **2006**, *10*, 61–66. [CrossRef]
12. Finestack, L.H.; Abbeduto, L. Expressive Language Profiles of Verbally Expressive Adolescents and Young Adults With Down Syndrome or Fragile X Syndrome. *J. Speech Lang. Heart Res.* **2010**, *53*, 1334–1348. [CrossRef]
13. DiGuiseppi, C.; Hepburn, S.; Davis, J.M.; Fidler, D.J.; Hartway, S.; Lee, N.; Miller, L.; Ruttenber, M.; Robinson, C. Screening for Autism Spectrum Disorders in Children With Down Syndrome. *J. Dev. Behav. Pediatr.* **2010**, *31*, 181–191. [CrossRef]
14. Oxelgren, U.W.; Myrelid, Å.; Annerén, G.; Ekstam, B.; Göransson, C.; Holmbom, A.; Isaksson, A.; Åberg, M.; Gustafsson, J.; Fernell, E. Prevalence of autism and attention-deficit-hyperactivity disorder in Down syndrome: A population-based study. *Dev. Med. Child Neurol.* **2016**, *59*, 276–283. [CrossRef]
15. Warner, G.; Howlin, P.; Salomone, E.; Moss, J.; Charman, T. Profiles of children with Down syndrome who meet screening criteria for autism spectrum disorder (ASD): A comparison with children diagnosed with ASD attending specialist schools. *J. Intellect. Disabil. Res.* **2016**, *61*, 75–82. [CrossRef]
16. Maenner, M.J.; Shaw, K.A.; Baio, J.; Washington, A.; Patrick, M.; DiRienzo, M.; Christensen, D.L.; Wiggins, L.D.; Pettygrove, S.; Andrews, J.G.; et al. Prevalence of Autism Spectrum Disorder Among Children Aged 8 Years—Autism and Developmental Disabilities Monitoring Network, 11 Sites, United States, 2016. *MMWR. Surveill. Summ.* **2020**, *69*, 1–12. [CrossRef] [PubMed]
17. Hamner, T.; Hepburn, S.; Zhang, F.; Fidler, D.; Rosenberg, C.R.; Robins, D.L.; Lee, N.R. Cognitive Profiles and Autism Symptoms in Comorbid Down Syndrome and Autism Spectrum Disorder. *J. Dev. Behav. Pediatr.* **2020**, *41*, 172–179. [CrossRef] [PubMed]
18. Godfrey, M.; Hepburn, S.; Fidler, D.J.; Tapera, T.; Zhang, F.; Rosenberg, C.R.; Lee, N.R. Autism spectrum disorder (ASD) symptom profiles of children with comorbid Down syndrome (DS) and ASD: A comparison with children with DS-only and ASD-only. *Res. Dev. Disabil.* **2019**, *89*, 83–93. [CrossRef] [PubMed]
19. Molloy, C.A.; Murray, D.S.; Kinsman, A.; Castillo, H.; Mitchell, T.; Hickey, F.J.; Patterson, B. Differences in the clinical presentation of Trisomy 21 with and without autism. *J. Intellect. Disabil. Res.* **2009**, *53*, 143–151. [CrossRef] [PubMed]
20. Moss, J.; Richards, C.; Nelson, L.; Oliver, C. Prevalence of autism spectrum disorder symptomatology and related behavioural characteristics in individuals with Down syndrome. *Autism* **2013**, *17*, 390–404. [CrossRef]

21. Capone, G.T.; Grados, M.A.; Kaufmann, W.E.; Bernad-Ripoll, S.; Jewell, A. Down syndrome and comorbid autism-spectrum disorder: Characterization using the aberrant behavior checklist. *Am. J. Med. Genet. Part A* **2005**, *134A*, 373–380. [CrossRef] [PubMed]

22. Ji, N.Y.; Capone, G.T.; Kaufmann, W.E. Autism spectrum disorder in Down syndrome: Cluster analysis of Aberrant Behaviour Checklist data supports diagnosis. *J. Intellect. Disabil. Res.* **2011**, *55*, 1064–1077. [CrossRef]

23. Abbeduto, L.; Berry-Kravis, E.; Sterling, A.; Sherman, S.; Edgin, J.O.; McDuffie, A.; Hoffmann, A.; Hamilton, D.; Nelson, M.; Aschkenasy, J.; et al. Expressive language sampling as a source of outcome measures for treatment studies in fragile X syndrome: Feasibility, practice effects, test-retest reliability, and construct validity. *J. Neurodev. Disord.* **2020**, *12*, 1–23. [CrossRef]

24. Thurman, A.J.; Edgin, J.O.; Sherman, S.L.; Sterling, A.; McDuffie, A.; Berry-Kravis, E.; Hamilton, D.; Abbeduto, L. Spoken language outcome measures for treatment studies in Down syndrome: Feasibility, practice effects, test-retest reliability, and construct validity of variables generated from expressive language sampling. *J. Neurodev. Disord.* **2021**, *13*, 1–17. [CrossRef] [PubMed]

25. Sansone, S.M.; Schneider, A.; Bickel, E.; Berry-Kravis, E.; Prescott, C.; Hessl, D. Improving IQ measurement in intellectual disabilities using true deviation from population norms. *J. Neurodev. Disord.* **2014**, *6*, 16. [CrossRef] [PubMed]

26. Roid, G. *Stanford Binet Intelligence Scales*, 5th ed.; Riverside Publishing: Rolling Meadows, IL, USA, 2003.

27. Lord, C.; Rutter, M.; DiLavore, P.; Risi, S.; Gotham, K.; Bishop, S.L. *Autism Diagnostic Observation Schedule-Second Edition (ADOS-2)*; Western Psychological Services: Los Angeles, CA, USA, 2012.

28. Gotham, K.; Pickles, A.; Lord, C. Standardizing ADOS Scores for a Measure of Severity in Autism Spectrum Disorders. *J. Autism Dev. Disord.* **2008**, *39*, 693–705. [CrossRef]

29. Hus, V.; Lord, C. The Autism Diagnostic Observation Schedule, Module 4: Revised Algorithm and Standardized Severity Scores. *J. Autism Dev. Disord.* **2014**, *44*, 1996–2012. [CrossRef] [PubMed]

30. Miller, J.; Chapman, R. *Systematic Analysis of Language Transcripts (SALT)*, Research version; SALT Software LLC: Middleton, WI, USA, 2008.

31. Benjamini, Y.; Hochberg, Y. Controlling the False Discovery Rate: A Practical and Powerful Approach to Multiple Testing. *J. R. Stat. Soc. Ser. B* **1995**, *57*, 289–300. [CrossRef]

32. Glennon, J.M.; D'Souza, H.; Mason, L.; Karmiloff-Smith, A.; Thomas, M.S. Visuo-attentional correlates of Autism Spectrum Disorder (ASD) in children with Down syndrome: A comparative study with children with idiopathic ASD. *Res. Dev. Disabil.* **2020**, *104*, 103678. [CrossRef]

33. Hepburn, S.L.; Moody, E.J. Diagnosing Autism in Individuals with Known Genetic Syndromes: Clinical Considerations and Implications for Intervention. In *International Review of Research in Developmental Disabilities*; Academic Press: Cambridge, MA, USA, 2011; Volume 40, pp. 229–259. [CrossRef]

34. Gotham, K.; Risi, S.; Pickles, A.; Lord, C. The Autism Diagnostic Observation Schedule: Revised Algorithms for Improved Diagnostic Validity. *J. Autism Dev. Disord.* **2007**, *37*, 613–627. [CrossRef]

35. Gotham, K.; Risi, S.; Dawson, G.; Tager-Flusberg, H.; Joseph, R.; Carter, A.; Hepburn, S.; McMahon, W.; Rodier, P.; Hyman, S.L.; et al. A Replication of the Autism Diagnostic Observation Schedule (ADOS) Revised Algorithms. *J. Am. Acad. Child Adolesc. Psychiatry* **2008**, *47*, 642–651. [CrossRef] [PubMed]

36. Bal, V.H.; Kim, S.H.; Fok, M.; Lord, C. Autism spectrum disorder symptoms from ages 2 to 19 years: Implications for diagnosing adolescents and young adults. *Autism Res.* **2019**, *12*, 89–99. [CrossRef] [PubMed]

37. Smith, E.; Næss, K.-A.B.; Jarrold, C. Assessing pragmatic communication in children with Down syndrome. *J. Commun. Disord.* **2017**, *68*, 10–23. [CrossRef] [PubMed]

38. Walden, T.A.; Blackford, J.U.; Carpenter, K.L. Differences in social signals produced by children with developmental delays of differing etiologies. *Am. J. Ment. Retard.* **1997**, *102*, 292–305. [CrossRef]

39. Fidler, D.J.; Most, D.E.; Booth-LaForce, C.; Kelly, J.F. Emerging Social Strengths in Young Children with Down Syndrome. *Infants Young-Child.* **2008**, *21*, 207–220. [CrossRef]

40. Hahn, L.J.; Loveall, S.J.; Savoy, M.T.; Neumann, A.M.; Ikuta, T. Joint attention in Down syndrome: A meta-analysis. *Res. Dev. Disabil.* **2018**, *78*, 89–102. [CrossRef]

41. Paparella, T.; Goods, K.S.; Freeman, S.; Kasari, C. The emergence of nonverbal joint attention and requesting skills in young children with autism. *J. Commun. Disord.* **2011**, *44*, 569–583. [CrossRef]

42. Abbeduto, L.; McDuffie, A.; Thurman, A.J. The fragile X syndrome–autism comorbidity: What do we really know? *Front. Genet.* **2014**, *5*, 355. [CrossRef]

43. McDuffie, A.; Thurman, A.J.; Hagerman, R.J.; Abbeduto, L. Symptoms of Autism in Males with Fragile X Syndrome: A Comparison to Nonsyndromic ASD Using Current ADI-R Scores. *J. Autism Dev. Disord.* **2015**, *45*, 1925–1937. [CrossRef]

44. Thurman, A.J.; McDuffie, A.; Kover, S.T.; Hagerman, R.J.; Abbeduto, L. Autism Symptomatology in Boys with Fragile X Syndrome: A Cross Sectional Developmental Trajectories Comparison with Nonsyndromic Autism Spectrum Disorder. *J. Autism Dev. Disord.* **2015**, *45*, 2816–2832. [CrossRef]

45. Bradbury, K.R.; Anderberg, E.I.; Huang-Storms, L.; Vasile, I.; Greene, R.K.; Duvall, S.W. Co-occurring Down Syndrome and Autism Spectrum Disorder: Cognitive, Adaptive, and Behavioral Characteristics. *J. Autism Dev. Disord.* **2021**, *51*, 1–12. [CrossRef]

brain
sciences

Article

Psychometric Evaluation of Social Cognition and Behavior Measures in Children and Adolescents with Down Syndrome

Emily K. Schworer [1,*], Emily K. Hoffman [1] and Anna J. Esbensen [1,2]

[1] Division of Developmental and Behavioral Pediatrics, Cincinnati Children's Hospital Medical Center, Cincinnati, OH 45229, USA; Emily.Hoffman1@cchmc.org (E.K.H.); Anna.Esbensen@cchmc.org (A.J.E.)
[2] Department of Pediatrics, University of Cincinnati College of Medicine, Cincinnati, OH 45267, USA
* Correspondence: Emily.Schworer@cchmc.org

Abstract: Individuals with Down syndrome (DS) are often described as socially engaged; however, challenges with social cognition, expressive language, and social interaction are also common in DS and are prospective outcomes of interest for clinical trials. The current study evaluates the psychometric properties of standardized measurements of social cognition and social behavior for potential use as outcome measures for children and adolescents with DS. Seventy-three youth ages 6 to 17 years old (M = 12.67, SD = 3.16) with DS were assessed on social cognition subtests of a neuropsychological assessment at two time points. Caregivers also completed a parent-report measure of social behavior. Measures were evaluated for feasibility, test-retest reliability, practice effects, convergent validity, and associations with broader developmental domains (i.e., age, cognition, and language). All social cognition and behavior measures met criteria for a portion of the psychometric indices evaluated, yet feasibility limitations were identified for the Developmental Neuropsychological Assessment, Second Edition (NEPSY-II) Affect Recognition subtest, and the NEPSY-II Theory of Mind subtest had problematic floor effects for percentile ranks. The Social Responsiveness Scale, Second Edition (SRS-2; T-scores) had high feasibility, moderate to excellent test-retest reliability, and no practice effects, suggesting this measure could be appropriate for use in clinical trials involving youth with DS.

Keywords: Down syndrome; social cognition; social behavior; measurement; children

Citation: Schworer, E.K.; Hoffman, E.K.; Esbensen, A.J. Psychometric Evaluation of Social Cognition and Behavior Measures in Children and Adolescents with Down Syndrome. *Brain Sci.* **2021**, *11*, 836. https://doi.org/10.3390/brainsci11070836

Academic Editors: James Kilner and Antonio Narzisi

Received: 17 May 2021
Accepted: 22 June 2021
Published: 24 June 2021

Publisher's Note: MDPI stays neutral with regard to jurisdictional claims in published maps and institutional affiliations.

1. Introduction

Individuals with Down syndrome (DS) are commonly described as socially engaged [1] and as having relatively strong nonverbal social functioning in early childhood [2]. Nevertheless, individuals with DS also experience challenges with core aspects of social relatedness including social cognition, expressive language, and social interaction [3–5]. Social challenges are further evident in rates of co-occurring autism spectrum disorder (ASD) in DS. Recent studies show that approximately 15–18% of children with DS also have an ASD diagnosis, which is markedly higher than the 1% reported in the general population [6,7]. Children with DS, with and without co-occurring ASD, experience social challenges that impede interactions with peers [5,8]. These social difficulties lead to greater potential for social isolation that, in turn, impacts mental health outcomes for this population [9], making social cognition and other social skills potential targets for intervention.

Social cognition is defined as the understanding of other's intentions, emotions, and behaviors [9–11]. This includes concepts such as theory of mind, which is the ability to reason about another's point of view, and affect recognition, the ability to identify emotions in others. Social cognition requires individuals to process and interpret social cues, and these skills impact the selection of social responses and subsequent quality of interactions with others in social contexts [12]. In children with ASD, specific connections have been made between social information processing and social behavior [12,13]. In children with DS, theory of mind performance is a greater relative challenge compared with children with other neurogenetic syndromes and intellectual disabilities, and their performance falls below their overall nonverbal cognitive abilities [14,15]. Affect and emotion recognition is also an area of challenge in DS in comparison with children with typical development matched on cognitive or receptive language level [16,17]. Studies using the Social Responsiveness Scale, Second Edition (SRS-2) [18] describe social communication and interactions in individuals with DS at low risk for ASD and show that these individuals have relative strengths in social motivation and challenges with social cognition, communication, and awareness [5,19]. For a review of social cognition development in DS, see [20].

A variety of measures have been used in past social cognition research in DS. Most of these measures are laboratory-based and include false belief tasks involving the location of objects [15,21,22] or the content of a container [15], appearance reality tasks [15,22], and emotion-matching tasks [16,17,23]. Although the majority of research on social cognition is completed with toddlers and preschool-aged children [9], these measures are used in the assessment of older children, adolescents, and young adults with DS [14,15]. Beyond the use of laboratory-based measures, standardized clinical assessments of social cognition have been used successfully to describe performance in other clinical populations such as ASD and Attention-Deficit/Hyperactivity Disorder (ADHD) [24,25]. However, standardized clinical assessments have yet to be evaluated to assess social cognition in DS. Another measure used to assess social cognition, among other social behaviors, is the SRS-2, and previous work supports its utility in 6- to 21-year-olds with DS [5,19]. A benefit of using the SRS-2 is that the parent reports on the child's typical social behavior without the child having to do more intensive in-person assessments. The SRS-2 is reported to have high internal consistency and concurrent validity with other ASD screeners among children with DS [19]; however, broader examination of test-retest reliability, practice effects, and convergent validity with direct assessments of social cognition has yet to be studied.

As we learn more about the social phenotype of children with DS and DS+ASD and as social challenges are better characterized [8,19], additional socially focused interventions tailored to children with DS will be needed. Pilot interventions targeting theory of mind skills have recently been completed with children and adolescents with DS [26] and suggest that these skills can be improved with targeted behavioral intervention. Because of the prospective growth of studies focused on social cognition and interaction in DS, a necessary first step to intervention studies is to validate social cognition and social behavior measures for this population.

Further, the priority to evaluate outcome measures for interventions and clinical trials in DS was expressed by the 2015 National Institutes of Health Down Syndrome Outcome Measure working group [27,28]. A summary from this working group identified no direct assessments of social cognition with evidence for use in DS but did state that the SRS-2 showed promise based on the sensitivity of the measure to detect ASD symptoms in DS [19,28]. Social cognition measures have been psychometrically evaluated in the general population [29–31]; however, continued efforts are needed to determine appropriate measures for DS. Psychometrically evaluating social cognition and social behavior measures in DS will ensure that assessments of these domains are suitable for children with DS and that there are no unintended floor effects due to the behavioral phenotype associated with DS. This psychometric validation is especially important for these measures as previous studies report expressive language artifacts in assessments of social cognition [21].

Present Study

The current study aimed to evaluate social cognition and social behavior measures in children and adolescents with DS. Evaluated measures included the Developmental Neuropsychological Assessment, Second Edition (NEPSY-II) social cognition subtests and the SRS-2. The first aim of the study was to quantify the number of participants with DS who were able to obtain scores on the measures (i.e., feasibility). Score distributions were also examined to determine if there were floor or ceiling effects. The second aim evaluated test-retest reliability, practice effects, convergent validity, and associations with broader developmental domains (age, cognition, and language). Lastly, additional investigation was completed to determine what ages and cognitive levels were appropriate for administration of any subtest with low feasibility. Psychometric evaluation of social cognition and behavior measures will improve the quality of measurement in DS research and inform future clinical trials.

2. Materials and Methods

2.1. Participants

Participants were 6- to 17-year-old children and adolescents with DS (n = 73; M chronological age = 12.67, SD = 3.16). Average IQ was 48.70, SD = 4.76, and deviation scores were used for all analyses (M = 33.79, SD = 13.75; described below) [32]. There was an approximately equal ratio of males and females (54.8% male). Most participants were White (87.7%) and non-Hispanic (93.2%). Two parents reported that their child had ASD. Data from the participants in this study have been used in other manuscripts focused on the assessment of working memory outcome measures [33] and the association between executive function and adaptive skills [34].

2.2. Procedure

All study procedures were approved by the Streamlined, Multisite, Accelerated Resources for Trials (SMART) IRB platform at Cincinnati Children's Hospital Medical Center (2018-0253, approved 23 April 2018), and informed consent was obtained for each subject before they participated. To be eligible for the study, participants were required to have a diagnosis of DS, have English as their primary language, and have an estimated nonverbal cognitive level of approximately three or older, per parent report, to support completion of at least a portion of the study procedures. Participants were recruited through medical clinics and DS associations at two sites. After being enrolled in the study, participants completed two visits, two weeks apart, as part of a broader longitudinal study on cognitive measurement in DS. To be included in analyses, participants were required to complete study measures at both time points.

2.3. Measures

2.3.1. Overall Cognitive Ability

Stanford-Binet, Fifth Edition (SB-5) [35]. Overall IQ was measured using the abbreviated battery IQ (ABIQ) to describe the cognitive abilities of the sample and examine associations with social measures. The SB-5 is a standardized measure of cognition and the ABIQ includes one nonverbal (Fluid Reasoning) and one verbal (Knowledge) subtest. The ABIQ is strongly correlated with the full-scale IQ in samples of children with a neurodevelopmental disorder [36], and reliability is also high for the ABIQ (r = 0.85–0.96) [35]. Deviation scores were used in this study to eliminate floor effects (deviation scoring procedures described in [32]). The ABIQ deviation scores are an estimate of the full-scale deviation scores. Negative scores are possible using deviation scoring and represent raw scores that for a participant's age are more than 3.33 standard deviations below the mean [32].

2.3.2. Language

Expressive Vocabulary Test, Third Edition (EVT-3) [37]. The EVT-3 is an expressive vocabulary measure and is designed for individuals 2.5–90+ years old. Participants were shown a picture and asked to label the picture or provide a synonym using a one-word response. Standard scores were used for all analyses. Three participants were unable to complete the measure because of low verbal ability.

Peabody Picture Vocabulary Test, Fifth Edition (PPVT-5) [38]. The PPVT-5 is a receptive vocabulary measure and is designed for individuals 2.5–90+ years old. Participants were shown four response options and were required to select the picture that was compatible with the word provided by the examiner. Standard scores were used for all analyses.

2.3.3. Social Cognition

Developmental Neuropsychological Assessment, Second Edition (NEPSY-II) [30]. The Theory of Mind and Affect Recognition subtests are both included in the NEPSY-II Social Perception domain and were selected to assess social cognition. Both subtests are designed for children as young as three years old and norming for the NEPSY-II included a variety of special group studies and small samples of children from clinical populations (e.g., intellectual disability, ADHD, and ASD). In the clinical sample, the correlation between Theory of Mind and Affect Recognition subtests is moderate ($r = 0.53$) and expected, considering the different abilities tested within the broader domain of social cognition.

NEPSY-II Theory of Mind

This subtest has verbal and nonverbal components and is designed to measure participants' understanding of intention, deception, belief, emotion, and pretending. The perception of others' thoughts, ideas, and feelings is also assessed. In the verbal portion of the task, the participant listens to scenarios or is shown images. The examiner asks a question about the point of view from a character in the presented information. In the nonverbal portion of the task (i.e., contextual task), the participant is presented with pictures of a social context and required to select the answer from four options that represents the correct affect of one of the persons pictured. NEPSY-II Theory of Mind test-retest reliability is high ($r = 0.84$). Because standard scores are only available up to 6:11, percentile ranks were used in addition to raw scores for measure description. Percentile ranks are unavailable for the total score for children 6–6:11 and therefore only verbal score percentile rank is reported for the 6-year-old participants. Verbal and total scores percentile rank are both reported for children 7–17 years old.

NEPSY-II Affect Recognition

This subtest is a nonverbal measure of the participant's ability to identify emotions of children in photographs. Types of emotions presented include happy, sad, anger, fear, disgust, and neutral. The subtest has four subsets of tasks that vary in instruction and involve: (1) stating if two faces have the same affect, (2) selecting two faces with the same affect, (3) selecting the face that matches the affect of the face at the top of the page, and (4) choosing two faces from memory that match the affect of a previously shown face. In the third trial type, there are two items presented on each page, and the item not being administered is covered by the examiner to reduce distraction. The Affect Recognition subtest demonstrates adequate to good test-retest reliability ($r = 0.46–0.66$). Raw and standard scores were used for analyses.

2.3.4. Social Behavior

Social Responsiveness Scale, Second Edition (SRS-2) [18]. The SRS-2 measures challenges with social interactions and communication. There are five subscales of the SRS-2 including Social Awareness, Social Cognition, Social Communication, Social Motivation, and Restricted Interests/Repetitive Behavior that produce a total score. The current study used parent report on the School-Age Form (ages 4–18). Parents were asked to rate their child's behavior over the last six months. Internal consistency correlations for the SRS-2 are high in the publisher's norming sample, which includes children with and without ASD ($\alpha = 0.95–0.97$) [18]. Similarly, high internal consistency has also been reported in smaller samples of children with DS ($\alpha = 0.94–0.96$) [19]. SRS-2 T-scores have a mean of 50 and a standard deviation of 10. Because both raw scores and T-scores had comparable results and there were no problems with score distributions identified, T-scores were used for all analyses.

2.4. Analysis Plan

To support Aim 1, the feasibility was assessed for measures of social cognition and social behavior administered to, or regarding, children and adolescents with DS. Feasibility was specified as the percentage of participants who provided responses at Time 1 and Time 2. Feasibility criteria were set a priori and ≥80% was the selected parameter for acceptable feasibility for use of these measures in DS research. This selection was informed by previous work on the psychometrics of cognitive measurements in intellectual disability and DS [33,39]. Examiners recorded reasons for noncompletion, which consisted of not understanding the task, behavioral noncompliance, and verbal refusal. Noncompletion of the parent-report measure was from missing questions (i.e., did not complete both sides of paper form) or failure to return the questionnaire. Range of scores, skewness, and kurtosis were also examined to determine the normality of the score distributions and to evaluate if there were floor effects for raw or standard scores. Acceptable values for skewness were between −1 and 1 and were between −2 and 2 for kurtosis. Participants who completed the measure with the lowest possible score, and those who were unable to complete the measure at Time 1, were both included in the estimate of floor effects. Floor effects < 20% were considered appropriate for research.

To support Aim 2, further psychometric evaluation (test-retest reliability, practice effects, and validity) was completed over the two-week testing interval. Test-retest reliability was assessed using intraclass correlation coefficients (ICC). Descriptive categories for ICCs are poor (<0.50), moderate (0.50−0.74), good (0.75−0.90), or excellent (>0.90) [33,40], and a priori good or excellent classifications were deemed suitable. Paired samples *t*-tests were used to assess practice effects. Practice effects were presumed if scores at the two testing visits had a significance value less than 0.05 and Cohen's *d* effect size greater than 0.20. Convergent validity across a selection of measures (NEPSY-II subtests and SRS-2 Social Awareness and Social Cognition) and associations among all social measures was determined using bivariate Pearson correlations. Correlation coefficients ≥0.50 were deemed as acceptable for convergent validity. Associations with broader developmental domains (age, cognition, and language) were also evaluated, and significant correlations were expected.

The third aim of the study investigated measures with low feasibility using post hoc sensitivity and specificity analyses. Sensitivity probabilities estimate the likelihood that a participant with specific characteristics will be able to complete the measure. Specificity probabilities estimate the likelihood that a participant not included in the specified characteristics will be unable to complete the measure. These analyses were completed for any measure that did not meet study feasibility criteria, and suggestions for age and cognitive ability of participants for future administration were established (as per [33]). Benchmarks for sensitivity and specificity probabilities were selected based on age (8 and 10 years) and cognitive ability (ABIQ deviation scores ≥20, ≥30, ≥40, and ≥50). Lower bounds of chronological age in previous clinical trials in DS informed benchmark selection [41].

3. Results

3.1. Aim 1: Feasibility and Floor Effects

Feasibility and floor effect indices for raw scores, percentile ranks, and/or standard scores (as appropriate) of the NEPSY-II Theory of Mind, NEPSY-II Affect Recognition, and SRS-2 are presented in Table 1. Two of the three measures evaluated in this study met the a priori criterion for feasibility: the NEPSY-II Theory of Mind (86.3–87.7%) and the SRS-2 (87.7%). NEPSY-II Affect Recognition fell below acceptable criterion for feasibility (71.2%) and therefore was investigated for Aim 3 as part of the post hoc analysis for low feasibility measures. Reasons for missing the NEPSY-II Affect Recognition subtest included not understanding the task (17.8%), behavioral noncompliance (3.4%), and verbal refusal (0.7%). A small portion of participants only completed the subtest at one time point (6.9%). Additionally, 15.4% of participants who completed the measure were described as exhibiting "acquiescence", defined as selecting responses without considering each response option. Floor effects followed the same pattern for the NEPSY-II raw scores and SRS-2 T-scores, with acceptable levels of floor effects for the NEPSY-II Theory of Mind (15.1–19.2%) and the SRS-2 (12.3%), and unacceptable levels for the NEPSY-II Affect Recognition (28.8%). Floor effects for percentile rank on the NEPSY-II Theory of Mind and standard scores on the NEPSY-II

Affect Recognition were both below a priori criteria. Specifically, of the participants who could complete the measures, 95% had the lowest percentile rank (<2%) on the NEPSY-II Theory of Mind Verbal, 97% had the lowest percentile rank (<2%) on the NEPSY-II Theory of Mind Total, and 39% had the lowest standard score (1) on the NEPSY-II Affect Recognition. Table 1 presents floor effects that include those with the lowest score on each measure and those who were unable to complete the task.

Table 1. Social cognition performance and feasibility at Time 1, *n* = 73.

	Min	Max	Median	Skew	Kurtosis	Feasibility *n* (%)	*n* at Floor [a]
NEPSY-II subtests							
ToM Verbal Raw Score	0	18	5.5	0.74	0.15	64 (87.7%)	14/73
ToM Verbal Percentile Rank [b]	<2	11–25	-	-	-		70/73
ToM Total Raw Score	0	22	7	0.64	-0.01	63 (86.3%)	11/73
ToM Total Percentile Rank [b]	<2	2–5	-	-	-		69/70 [c]
AR Total Raw Score	1	26	13	−0.20	−0.57	52 (71.2%)	21/73
AR Total Standard Score	1	8	2	0.94	−0.32		41/73
SRS-2 T-scores							
Total	42	86	60.5	0.54	0.39	64 (87.7%)	9/73
Social Awareness	37	79	60	−0.18	−0.37		
Social Cognition	43	82	65	−0.08	−0.68		
Social Communication	41	90	60	0.83	1.35		
Social Motivation	40	90	54	1.10	1.64		
RRBI	45	90	57	0.65	−0.40		

[a] *n* at floor includes children who got the lowest score on the measure and those who could not complete the task; [b] Some descriptive statistics not reported for percentile ranks; [c] NEPSY-II Theory of Mind Total score percentile rank is not normed for 6-year-olds (*n* = 3); ToM = Theory of Mind; AR = Affect Recognition; SRS-2 = Social Responsiveness Scale, Second Edition; RRBI = Restricted, Repetitive Behaviors and Interests.

3.2. Aim 2: Test-Retest Reliability, Practice Effects, and Validity

3.2.1. Test-Retest Reliability and Practice Effects

Overall test-retest reliability ranged from poor to excellent on the evaluated measures (Table 2). Raw scores for the NEPSY-II Theory of Mind verbal and total scores were in the moderate range for test-retest reliability, falling below a priori criterion for this study. The NEPSY-II Affect Recognition had poor test-retest reliability, again below acceptable criterion. The SRS-2 had moderate to excellent test-retest reliability and all ICCs were 0.70 or greater. The majority were above 0.75 a priori criterion, therefore demonstrating stable test-retest reliability across the two-week testing interval. There were no practice effects on any of the measures evaluated in this study (see Table 2).

3.2.2. Convergent Validity and Associations among Social Measures

Convergent validity was assessed for a selection of measures (NEPSY-II subtests and SRS-2 Social Awareness and Social Cognition), and correlations among all social measures were examined (Table 3). Associations between NEPSY-II Theory of Mind and Affect Recognition were below the acceptable criterion of *r* > 0.50; however, the correlation coefficients of 0.43–0.51 were similar to data on the relation between the two measures reported by the publisher (*r* = 0.53) [30]. Significant associations were also found between SRS-2 Social Awareness and SRS-2 Social Cognition. However, there were no significant associations between the NEPSY-II subtests and the SRS-2 Social Awareness or Social Cognition. Although not all subdomains of the SRS-2 are theoretically aligned with the NEPSY-II direct assessments of social cognition, it is noteworthy that no SRS-2 subscales were correlated with NEPSY-II social cognition subtests. Within the SRS-2, all subscales were positively correlated, and the strength of many of the associations were strong (0.34–0.94).

Table 2. Examination of practice effects, test-retest reliability, and correlations with broader developmental domains.

	Time 1 Mean (SD)	Time 2 Mean (SD)	t	p	d	ICC	Age	ABIQ [a]	EVT-3 SS	PPVT-5 SS
NEPSY-II subtests										
ToM Verbal Raw Score	5.63 (4.24)	5.03 (3.85)	1.37	0.18	0.15	0.63	0.13	0.50 ***	0.42 ***	0.41 **
ToM Total Raw Score	8.19 (4.77)	7.92 (4.34)	0.57	0.57	0.06	0.66	0.15	0.51 ***	0.47 ***	0.44 ***
AR Total Raw Score	13.48 (6.80)	13.23 (6.76)	0.25	0.81	0.04	0.41	0.32 *	0.59 ***	0.39 **	0.25
SRS-2 T-scores										
Total	61.28 (9.67)	61.67 (10.87)	−0.78	0.44	0.04	0.92	0.08	−0.31 *	−0.08	−0.21
Social Awareness	59.47 (9.04)	59.72 (9.06)	−0.03	0.78	0.03	0.70	−0.08	−0.32 *	−0.19	−0.25 *
Social Cognition	63.64 (9.24)	64.34 (9.99)	−1.01	0.32	0.07	0.83	−0.02	−0.32 *	−0.15	−0.22
Social Communication	61.02 (9.46)	61.86 (11.34)	−1.24	0.22	0.08	0.86	0.06	−0.29 *	−0.05	−0.17
Social Motivation	54.25 (10.38)	53.86 (11.64)	0.59	0.56	0.04	0.89	0.11	−0.11	−0.19	−0.01
RRBI	60.16 (10.84)	60.36 (11.64)	−0.34	0.74	0.02	0.91	0.15	−0.27 *	−0.17	−0.24

* $p < 0.05$, ** $p < 0.01$, *** $p < 0.001$; [a] Stanford-Binet, Fifth Edition Deviation Scores; ABIQ = Abbreviated Intelligence Quotient; PPVT-5 = Peabody Picture Vocabulary Test, Fifth Edition; EVT-3 = Expressive Vocabulary Test, Third Edition; SS = Standard Score; SRS-2 = Social Responsiveness Scale, Second Edition; ICC = intraclass correlation coefficients; ToM = Theory of Mind; AR = Affect Recognition; RRBI = Restricted, Repetitive Behaviors and Interests; d = Cohen's d.

Table 3. Bivariate Pearson correlations to assess convergent validity and associations among social measures at Time 1.

	1	2	3	4	5	6	7	8
1. Theory of Mind Verbal Raw Score								
2. Theory of Mind Total Raw Score	0.96 ***							
3. Affect Recognition Total Raw Score	0.43 **	0.51 **						
4. SRS-2 Social Awareness	−0.14	−0.14	−0.21					
5. SRS-2 Social Cognition	−0.15	−0.15	−0.18	0.76 ***				
6. SRS-2 Social Communication	−0.05	−0.08	−0.04	0.69 ***	0.76 ***			
7. SRS-2 Social Motivation	0.09	0.07	0.11	0.34 **	0.56 ***	0.71 ***		
8. SRS-2 RRBI	−0.18	−0.14	−0.11	0.54 ***	0.60 ***	0.69 ***	0.64 ***	
9. SRS-2 Total	−0.10	−0.10	−0.08	0.74 ***	0.85 ***	0.94 ***	0.80 ***	0.83 ***

** $p < 0.01$; *** $p < 0.001$; SRS−2 = Social Responsiveness Scale, Second Edition; RRBI = Restricted, Repetitive Behaviors and Interests.

3.2.3. Associated Developmental Domains

Significant positive correlations were observed between NEPSY-II subtests and ABIQ deviation scores, EVT-3 standard scores, and PPVT-5 standard scores (Table 2). For the NEPSY-II Theory of Mind, associations with all three cognition and language measures were moderately strong (0.41–0.51). The NEPSY-II Affect Recognition subtest was also positively correlated with ABIQ deviation scores and EVT-3 standard scores; however, no association was found with PPVT-5 standard scores. The SRS-2 had modest correlations with ABIQ deviation scores in the expected direction, such that more social behavior challenges were associated with lower ABIQ. In most cases, there was no significant correlation between the SRS-2 and PPVT-5 or EVT-3 standard scores. The majority of the measures were not associated with chronological age, with the exception of the NEPSY-II Affect Recognition Total raw score, which was positively correlated with age ($r = 0.32$).

3.3. Aim 3: Assessments with Low Feasibility

The NEPSY-II Affect Recognition was the only measure to fall below the feasibility threshold in this study. To better understand the subset of the population within DS that this measure would be appropriate for, sensitivity and specificity calculations were completed (Table 4). Less restrictive guidelines (i.e., ABIQ deviation $\geq$ 20 or 30) provided higher sensitivity, indicating that completers of the measure were correctly identified. As guidelines become more restrictive (i.e., ABIQ deviation $\geq$ 40 or 50), sensitivity decreased, and not all participants who could complete the task were identified using the more limiting benchmarks. More restrictive ABIQ also led to higher specificity, indicating that those who were *not* able to complete the measure were correctly identified when using those more restrictive benchmarks. Chronological ages examined (8 and 10 years) revealed minimal differences between sensitivity and specificity probabilities. Figure 1 illustrates the chronological age and ABIQ deviation scores of both completers and non-completers for the NEPSY-II Affect Recognition in our sample.

Table 4. Post hoc sensitivity and specificity for the measure below feasibility criteria.

| | NEPSY-II Affect Recognition | | | |
| | Sensitivity | | Specificity | |
	Age 8	Age 10	Age 8	Age 10
No ABIQ [a] Restriction	94.2%	88.5%	23.8%	33.3%
ABIQ [a] ≥ 20	94.0%	88.2%	42.9%	52.4%
ABIQ [a] ≥ 30	74.0%	70.6%	90.5%	90.5%
ABIQ [a] ≥ 40	38.0%	38.0%	100%	100%
ABIQ [a] ≥ 50	14.0%	14.0%	100%	100%

ABIQ [a] = Stanford-Binet, Fifth Edition abbreviated battery IQ deviation score.

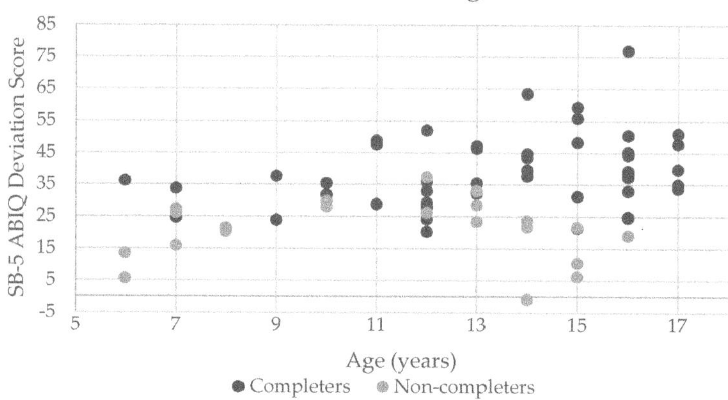

Figure 1. Chronological age in years and Stanford-Binet, Fifth Edition abbreviated battery IQ deviation scores of completers and non-completers for NEPSY-II Affect Recognition. Negative score represents a raw score more than 3.33 standard deviations below the mean that for the participant's age [32].

4. Discussion

This study evaluated the psychometric properties of two clinical assessments of social cognition and one social behavior parent questionnaire (summarized in Table 5). Both direct assessments of social cognition and parent-report of social behavior met criteria for a portion of the psychometric indices evaluated. Associations with cognition and language abilities emphasize how these social cognition and behavior measures relate to broader developmental domains. Additionally, the relations among measures show a clear pattern of correlation within NEPSY-II subtests and within SRS-2 subdomains, but no correlations were found across the direct assessments and parent-report measures. The SRS-2 demonstrated the strongest psychometric properties, with high feasibility, moderate to excellent test-retest reliability, and no practice effects, suggesting this measure could be appropriate for use in clinical trials involving youth with DS. The NEPSY-II Theory of Mind subtest raw scores also demonstrated good psychometrics; however, percentile rank floor effects indicate this measure is not suitable for this population. Feasibility for the NEPSY-II Affect Recognition was problematic and this measure may only be appropriate for certain IQ ranges of children and adolescents with DS.

Table 5. Summary of social cognition and behavior measures assessed on a priori criteria.

	Minimal Floor Effects	Feasibility	Test-Retest	Negligible Practice Effects
NEPSY-II				
Theory of Mind Verbal Raw Score	+	+	−	+
Theory of Mind Total Raw Score	+	+	−	+
Affect Recognition Total Raw Score	−	−	−	+
SRS-2 T scores				
Social Awareness	+	+	−	+
Social Cognition	+	+	+	+
Social Communication	+	+	+	+
Social Motivation	+	+	+	+
RRBI	+	+	+	+
Total	+	+	+	+

+ indicates study criterion met: <20% floor effects, $\geq$80% feasibility, $\geq$0.75 test-retest ICC, small and non-significant practice effects; − indicates study criterion not met: $\geq$20% floor effects, <80% feasibility, <0.75 test-retest ICC, medium/large and significant practice effects; RRBI = Restricted, Repetitive Behaviors and Interests.

4.1. Feasibility and Floor Effects

The NEPSY-II Theory of Mind and SRS-2 both met a priori feasibility criteria and over 85% of participants obtained scores on these measures. Although feasibility was adequate for these measures, percentile ranks were at the floor for the NEPSY-II Theory of Mind. Therefore, we recommend raw scores for use in future work utilizing this measure. Despite relative challenges with theory of mind in DS [14,15], it is encouraging that raw scores were able to capture a range of scores on this measure; however, percentile rank floor effects show that this measure does not discriminate performance between subjects in the sample using published norms. Further, there were minimal differences between raw scores and T-scores on the SRS-2, and the use of T-scores is appropriate for this measure of social behavior. High feasibility of the SRS-2 reinforces the suitability of this tool for the measurement of social behavior in individuals with DS [5,19]. Feasibility was below a priori criterion for the NEPSY-II Affect Recognition and floor effects were observed for both raw and standard scores and most problematic for standard scores. Low feasibility and standard scores on this measure may be, in part, due to difficulties individuals with DS have with recognizing emotional expressions in others [16,17]. Difficulty understanding the task was the greatest reason for noncompletion and assessments of affect recognition with simpler instructions and task demands may be needed for this population. Recommendations for future use of the NEPSY-II Affect Recognition in DS are provided in the discussion of low feasibility measures below.

4.2. Test-Retest Reliability, Practice Effects, and Validity

There were mixed results regarding the reliability and validity of the evaluated measures. Test-retest reliability was strongest for the SRS-2 subscales, providing evidence for consistent reports of social behavior by caregivers. Social Awareness had the lowest ICC for the SRS-2 (0.70), which was in the moderate range, but close to the "good" category (0.75). Although moderate test-retest reliability was found for NEPSY-II Theory of Mind raw scores, NEPSY-II Affect Recognition raw scores demonstrated poor reliability. Inconsistent scores between the two-week test-retest interval on the NEPSY-II Affect Recognition indicate that children may be guessing or acquiescing with their responses. For all evaluated measures, NEPSY-II subtests and SRS-2, practice effects were negligible. The lack of improvement at the second study time point suggests the measures were stable with multiple administrations over a relatively short period.

Investigation of convergent validity resulted in no association between parent reports of social cognition/awareness and direct assessments of social cognition. These different test modalities may be tapping different constructs or skills, as there are clear differences in laboratory-based assessments compared with parent-report measures. Therefore, while the NEPSY-II Theory of Mind shows some good psychometric properties, we need to consider what it is measuring. It may be the case that standardized clinical assessments of theory of mind do not represent parental perceptions of a child's daily abilities in social awareness and understanding. The NEPSY-II Theory of Mind may also have poor ecological validity. Further, the NEPSY-II Theory of Mind is moderately correlated with receptive and expressive language, and overall language abilities may be confounding performance, as has occurred in previous studies [21]. Another plausible interpretation is that social abilities reported by parents are truly different skills than those assessed in the laboratory. Although there were no associations between the NEPSY-II and SRS-2, there were significant associations among the SRS-2 subscales, which parallels previous significant correlations reported among SRS subscales in DS [19].

Associations with broader developmental domains varied significantly across measures. First, both NEPSY-II subtests and SRS-2 subscales had significant correlations with cognitive ability, in the expected directions, such that higher cognitive abilities were associated with better social cognition and fewer social behavior challenges. The associations between the SRS-2 subscales and cognitive abilities have not been consistently found in previous investigations between SRS-2 and nonverbal IQ in DS [5,19], but this study does replicate a moderate association found between cognition and SRS Total T-scores [19], despite using different IQ measures. Correlations between NEPSY-II subtests and ABIQ were markedly stronger than comparisons between the SRS-2 and ABIQ. This reinforces the idea that direct assessment may be tapping similar skills that are fundamentally different from the behaviors and performance observed by parents in the home environment. Both NEPSY-II subtests were positively correlated with the expressive language measures, but only the NEPSY-II Theory of Mind subtest was associated with receptive language. This highlights the receptive language demands of the NEPSY-II Theory of Mind that are required to complete the measure. SRS-2 Social Awareness was the only subscale that was associated with receptive language, which deviates from previous reports of a significant association between all SRS subscales and receptive vocabulary [19]. This study also replicated previous reports of no correlation between the SRS-2 subscales and expressive language [5]. Finally, associations with chronological age were minimal and corroborate previous reports of a lack of association with the SRS [5,19], suggesting that developmental level is a better indicator of social cognition and behavior than age.

4.3. Assessments with Low Feasibility

Because the NEPSY-II Affect Recognition had feasibility that fell below a priori criterion, follow-up post hoc sensitivity and specificity analyses were used to describe who the measure is appropriate for within the sample of children and adolescents with DS. There was a clear pattern that less restrictive guidelines led to more sensitivity, correctly identifying any participant who could complete the task. More restrictive guidelines led to more precision and greater confidence that those in the high IQ ranges could complete the measure (i.e., specificity). It is ideal to have a balance of both high sensitivity and specificity to avoid missing participants who could complete the task but also to be administering a task appropriate for the individuals in a study or clinical trial. The current study's benchmark of ABIQ deviation scores ≥ 30 had the greatest balance between sensitivity and specificity probabilities and would be appropriate for inclusion/exclusion criteria if the NEPSY-II Affect Recognition were a required measure for a testing battery. However, there are limitations to this benchmark, as there were some participants below the ABIQ deviation score of 30 who were able to complete the measure.

4.4. Limitations and Future Directions

The current study provides essential information about the psychometric properties of social cognition and social behavior measures in DS, but it also has limitations. First, the rates of ASD in our study sample were lower than what has been reported in other studies examining the prevalence of ASD in DS, and additional work is needed to determine if these measures are appropriate for participants with DS and co-occurring ASD. There is also a need for a longer follow-up period to determine how these measures assess social constructs over 6 months or a year, to match the study design of a clinical trial. Examining the psychometrics of the measures in groups of children within narrower age ranges will also be an important step for future research. Additionally, social behavior was only measured using parent-report, and while it is valuable to understand the comparison between direct assessment and parent-report, this study did not include any direct assessments of social behavior. Because the NEPSY-II laboratory-based assessments were not correlated with the SRS-2 Social Cognition and Social Awareness, additional work is also needed to determine the generalizability of NEPSY-II subtests to real-world contexts. Finally, because few standardized clinical assessments focus on social cognition, further examination of a greater variety of social cognition laboratory-based measures is needed to ensure that the measures appropriate for the general population [29,31] are also suitable for individuals with DS. This future work would help to identify additional alternatives for measuring social cognition and social behavior in DS.

5. Conclusions

Findings from this study add to the list of standardized measures that may be used in clinical trials with children and adolescents with DS. The SRS-2 T-scores had normal distributions, good feasibility, moderate to excellent test-retest reliability, and no practice effects, and therefore this measure could be suitable for use in clinical trials. Although the NEPSY-II Theory of Mind raw scores were psychometrically sound, the measure was problematic overall, considering the percentile rank floor effects and lack of evidence for ecological validity. Researchers should also use caution when using NEPSY-II Affect Recognition, as feasibility was problematic in the current study. We recommend referencing the sensitivity and specificity benchmarks when using this measure to guide decisions about inclusion/exclusion criteria in future studies with this population. The psychometric evaluation of social cognition and social behavior measures supports the NIH working group initiative of determining appropriate outcome measures for individuals with DS [27] and will contribute to the success of future clinical trials in DS.

Author Contributions: Conceptualization, E.K.S., E.K.H. and A.J.E.; methodology, E.K.S. and E.K.H.; formal analysis, E.K.S.; investigation, E.K.S., E.K.H. and A.J.E.; data curation, E.K.S., E.K.H. and A.J.E.; writing—original draft preparation, E.K.S.; writing—review and editing, E.K.S., E.K.H. and A.J.E.; visualization, E.K.S.; supervision, A.J.E.; project administration, E.K.H.; funding acquisition, A.J.E. All authors have read and agreed to the published version of the manuscript.

Funding: This manuscript was funded by the Eunice Kennedy Shriver National Institute of Child Health and Human Development of the National Institutes of Health, R01 HD093754.

Institutional Review Board Statement: The study was conducted according to the guidelines of the Declaration of Helsinki and was approved by the Institutional Review Board of Cincinnati Children's Hospital Medical Center (2018-0253, approved 4/23/2018).

Informed Consent Statement: Informed consent was obtained from all subjects involved in the study.

Data Availability Statement: The data that support the findings of this study are available from the corresponding author upon reasonable request.

Acknowledgments: This research would not have been possible without the contributions of the participating families and the community's support.

Conflicts of Interest: The authors declare no conflict of interest. The funders had no role in the design of the study; in the collection, analyses, or interpretation of data; in the writing of the manuscript; or in the decision to publish the results.

References

1. Daunhauer, L.A.; Fidler, D.J. The Down syndrome behavioral phenotype: Implications for practice and research in occupational therapy. *Occup. Ther. Health Care* **2011**, *25*, 7–25. [CrossRef] [PubMed]
2. Fidler, D.J.; Philofsky, A.; Hepburn, S.L.; Rogers, S.J. Nonverbal Requesting and Problem-Solving by Toddlers with Down Syndrome. *Am. J. Ment. Retard.* **2005**, *110*, 312–322. [CrossRef]
3. Chapman, R.S. (Ed.) *Language and Cognitive Development in Children and Adolescents with Down Syndrome*; Paul H. Brookes Publishing: Baltimore, MD, USA, 1999; pp. 41–60.
4. Abbeduto, L.; McDuffie, A.; Thurman, A.; Kover, S. Language development in individuals with intellectual and developmental disabilities: From phenotypes to treatments. In *International Review of Research in Developmental Disabilities*; Elsevier: Amsterdam, The Netherlands, 2016; Volume 50, pp. 71–118.
5. Channell, M.M. The Social Responsiveness Scale (SRS-2) in school-age children with Down syndrome at low risk for autism spectrum disorder. *Autism Dev. Lang. Impair.* **2020**, *5*, 2396941520962406. [CrossRef]
6. DiGuiseppi, C.; Hepburn, S.; Davis, J.M.; Fidler, D.J.; Hartway, S.; Lee, N.R.; Miller, L.; Ruttenber, M.; Robinson, C. Screening for autism spectrum disorders in children with Down syndrome: Population prevalence and screening test characteristics. *J. Dev. Behav. Pediatrics* **2010**, *31*, 181–191. [CrossRef] [PubMed]
7. Hepburn, S.; Philofsky, A.; Fidler, D.J.; Rogers, S. Autism symptoms in toddlers with Down syndrome: A descriptive study. *J. Appl. Res. Intellect. Disabil.* **2008**, *21*, 48–57. [CrossRef]
8. Hamner, T.; Hepburn, S.; Zhang, F.; Fidler, D.; Rosenberg, C.R.; Robins, D.L.; Lee, N.R. Cognitive Profiles and Autism Symptoms in Comorbid Down Syndrome and Autism Spectrum Disorder. *J. Dev. Behav. Pediatrics* **2020**, *41*, 172–179. [CrossRef]
9. Cebula, K.R.; Moore, D.G.; Wishart, J.G. Social cognition in children with Down's syndrome: Challenges to research and theory building. *J. Intellect. Disabil. Res.* **2010**, *54*, 113–134. [CrossRef]
10. Happé, F.; Frith, U. Annual research review: Towards a developmental neuroscience of atypical social cognition. *J. Child Psychol. Psychiatry Allied Discip.* **2014**, *55*, 553–557. [CrossRef]
11. Garfield, J.; Peterson, C.; Perry, T. Social Cognition, Language Acquisition and The Development of the Theory of Mind. *Mind Lang.* **2001**, *16*, 494–541. [CrossRef]
12. Pino, M.C.; Mariano, M.; Peretti, S.; D'Amico, S.; Masedu, F.; Valenti, M.; Mazza, M. When do children with autism develop adequate social behaviour? Cross-sectional analysis of developmental trajectories. *Eur. J. Dev. Psychol.* **2020**, *17*, 71–87. [CrossRef]
13. Ziv, Y.; Hadad, B.S.; Khateeb, Y. Social information processing in preschool children diagnosed with autism spectrum disorder. *J. Autism Dev. Disord.* **2014**, *44*, 846–859. [CrossRef] [PubMed]
14. Abbeduto, L.; Pavetto, M.; Kesin, E.; Weissman, M.D.; Karadottir, S.; O'Brien, A.; Cawthon, S. The linguistic and cognitive profile of Down syndrome: Evidence from a comparison with fragile X syndrome. *Downs Syndr. Res. Pract.* **2001**, *7*, 9–15. [CrossRef]
15. Giaouri, S.; Alevriadou, A.; Tsakiridou, E. Theory of mind abilities in children with Down syndrome and non-specific intellectual disabilities: An empirical study with some educational implications. *Procedia Soc. Behav. Sci.* **2010**, *2*, 3883–3887. [CrossRef]
16. Williams, K.R.; Wishart, J.G.; Pitcairn, T.K.; Willis, D.S. Emotion recognition by children with Down syndrome: Investigation of specific impairments and error patterns. *Am. J. Ment. Retard.* **2005**, *110*, 378–392. [CrossRef]
17. Hippolyte, L.; Barisnikov, K.; Van der Linden, M.; Detraux, J.J. From facial emotional recognition abilities to emotional attribution: A study in Down syndrome. *Res. Dev. Disabil.* **2009**, *30*, 1007–1022. [CrossRef] [PubMed]
18. Constantino, J.N.; Gruber, C.P. *Social Responsiveness Scale*, 2nd ed.; Western Psychological Services: Torrance, CA, USA, 2012.
19. Channell, M.M.; Phillips, B.A.; Loveall, S.J.; Conners, F.A.; Bussanich, P.M.; Klinger, L.G. Patterns of autism spectrum symptomatology in individuals with Down syndrome without comorbid autism spectrum disorder. *J. Neurodev. Disord.* **2015**, *7*, 5. [CrossRef] [PubMed]
20. Cebula, K.R.; Wishart, J.G. Social Cognition in Children with Down Syndrome. In *International Review of Research in Mental Retardation*; Academic Press: Cambridge, MA, USA, 2008; Volume 35, pp. 43–86.
21. Abbeduto, L.; Short-Meyerson, K.; Benson, G.; Dolish, J. Relationship between theory of mind and language ability in children and adolescents with intellectual disability. *J. Intellect. Disabil. Res.* **2004**, *48*, 150–159. [CrossRef] [PubMed]
22. Cornish, K.; Burack, J.; Rahman, A.; Munir, F.; Russo, N.; Grant, C. Theory of mind deficits in children with fragile X syndrome. *J. Intellect. Disabil. Res.* **2005**, *49*, 372–378. [CrossRef]
23. Wishart, J.G.; Cebula, K.R.; Willis, D.S.; Pitcairn, T.K. Understanding of facial expressions of emotion by children with intellectual disabilities of differing aetiology. *J. Intellect. Disabil. Res.* **2007**, *51*, 551–563. [CrossRef]
24. Miranda, A.; Berenguer, C.; Roselló, B.; Baixauli, I.; Colomer, C. Social Cognition in Children with High-Functioning Autism Spectrum Disorder and Attention-Deficit/Hyperactivity Disorder. Associations with Executive Functions. *Front. Psychol.* **2017**, *8*, 1035. [CrossRef]
25. Loukusa, S.; Mäkinen, L.; Kuusikko-Gauffin, S.; Ebeling, H.; Moilanen, I. Theory of mind and emotion recognition skills in children with specific language impairment, autism spectrum disorder and typical development: Group differences and

connection to knowledge of grammatical morphology, word-finding abilities and verbal working memory. *Int. J. Lang. Commun. Disord.* **2014**, *49*, 498–507. [CrossRef] [PubMed]

26. Comblain, A.; Schmetz, C. Improving Theory of Mind skills in Down Syndrome? A Pilot Study. *J. Cogn. Educ. Psychol.* **2020**, *19*. [CrossRef]
27. Eunice Kennedy Shriver National Institute of Child Health and Human Development. Outcome Measures for Clinical Trials in Individuals with Down Syndrome. 2015. Available online: https://www.nichd.nih.gov/about/meetings/2015/Documents/DS_outcomes_meeting_summary.pdf (accessed on 17 August 2020).
28. Esbensen, A.J.; Hooper, S.R.; Fidler, D.; Hartley, S.L.; Edgin, J.; d'Ardhuy, X.L.; Capone, G.; Conners, F.A.; Mervis, C.B.; Abbeduto, L.; et al. Outcome measures for clinical trials in Down syndrome. *Am. J. Intellect. Dev. Disabil.* **2017**, *122*, 247–281. [CrossRef] [PubMed]
29. Pinkham, A.E.; Penn, D.L.; Green, M.F.; Harvey, P.D. Social Cognition Psychometric Evaluation: Results of the Initial Psychometric Study. *Schizophr. Bull.* **2016**, *42*, 494–504. [CrossRef]
30. Korkman, M.; Kirk, U.; Kemp, S. *NEPSY Second Edition (NEPSY-II)*; Harcourt Assessment: San Antonio, TX, USA, 2007.
31. Pinkham, A.E.; Harvey, P.D.; Penn, D.L. Social Cognition Psychometric Evaluation: Results of the Final Validation Study. *Schizophr. Bull.* **2018**, *44*, 737–748. [CrossRef]
32. Sansone, S.M.; Schneider, A.; Bickel, E.; Berry-Kravis, E.; Prescott, C.; Hessl, D. Improving IQ measurement in intellectual disabilities using true deviation from population norms. *J. Neurodev. Disord.* **2014**, *6*, 16. [CrossRef]
33. Schworer, E.K.; Esbensen, A.J.; Fidler, D.J.; Beebe, D.W.; Carle, A.; Wiley, S. Evaluating working memory outcome measures for children with Down syndrome. *J. Intellect. Disabil. Res.* **2021**. [CrossRef]
34. Will, E.A.; Schworer, E.K.; Esbensen, A.J. The role of distinct executive functions on adaptive behavior in children and adolescents with Down syndrome. *Child Neuropsychol.* **2021**, 1–19. [CrossRef] [PubMed]
35. Roid, G. *Stanford-Binet Intelligence Scale: Fifth Edition*; Riverside: Chicago, CA, USA, 2003.
36. Twomey, C.; O'Connell, H.; Lillis, M.; Tarpey, S.L.; O'Reilly, G. Utility of an abbreviated version of the stanford-binet intelligence scales in estimating 'full scale' IQ for young children with autism spectrum disorder. *Autism Res.* **2018**, *11*, 503–508. [CrossRef] [PubMed]
37. Williams, K.T. *Expressive Vocabulary Test*, 3rd ed.; Pearson: Bloomington, MN, USA, 2018.
38. Dunn, D.M. *Peabody Picture Vocabulary Test*, 5th ed.; Pearson: Bloomington, MN, USA, 2019.
39. Hessl, D.; Sansone, S.M.; Berry-Kravis, E.; Riley, K.; Widaman, K.F.; Abbeduto, L.; Schneider, A.; Coleman, J.; Oaklander, D.; Rhodes, K.C. The NIH Toolbox Cognitive Battery for intellectual disabilities: Three preliminary studies and future directions. *J. Neurodev. Disord.* **2016**, *8*, 35. [CrossRef]
40. Koo, T.K.; Li, M.Y. A Guideline of Selecting and Reporting Intraclass Correlation Coefficients for Reliability Research. *J. Chiropr. Med.* **2016**, *15*, 155–163. [CrossRef] [PubMed]
41. Kishnani, P.S.; Heller, J.H.; Spiridigliozzi, G.A.; Lott, I.; Escobar, L.; Richardson, S.; Zhang, R.; McRae, T. Donepezil for treatment of cognitive dysfunction in children with Down syndrome aged 10–17. *Am. J. Med Genet. Part A* **2010**, *152*, 3028–3035. [CrossRef] [PubMed]

brain sciences

Article

Description of Daily Living Skills and Independence: A Cohort from a Multidisciplinary Down Syndrome Clinic

Kavita Krell [1,*], Kelsey Haugen [1], Amy Torres [1] and Stephanie L. Santoro [1,2]

1   Division of Medical Genetics, Department of Pediatrics, Massachusetts General Hospital, Boston, MA 02114, USA; haugenkr@miamioh.edu (K.H.); aetorres@partners.org (A.T.); ssantoro3@mgh.harvard.edu (S.L.S.)
2   Department of Pediatrics, Harvard Medical School, Boston, MA 02115, USA
*   Correspondence: kkrell@mgh.harvard.edu

Abstract: Levels of independence vary in individuals with Down syndrome (DS). We began this study to describe the current life skills in our clinic population of children and adults with DS. We collected and reviewed demographics, living situation, and life skills from an electronic intake form used in clinic procedures. Descriptive statistics for this cohort study included mean, standard deviation, and frequencies. From 2014–2020, 350 pediatric and 196 adult patients (range 0–62 years) with a first visit to the Massachusetts General Hospital Down Syndrome Program are described. Pediatric patients were most often enrolled in school, and in an inclusion setting. Adult patients were most often participating in a day program, living with family, and wanted to continue living with family in the future. Most (87%) of adults with DS communicated verbally, though fewer could use written communication (17%). Life skills of greatest importance to adolescents and adults with DS included: learning about healthy foods (35%), preparing meals (34%), providing personal information when needed (35%), and describing symptoms to a doctor (35%). Life skills for patients with DS are varied; those associated with a medical appointment, such as sharing symptoms with the doctor, could improve for greater independence.

Keywords: Trisomy 21; Down syndrome; independence; transition to adulthood; proxy-report

Citation: Krell, K.; Haugen, K.; Torres, A.; Santoro, S.L. Description of Daily Living Skills and Independence: A Cohort from a Multidisciplinary Down Syndrome Clinic. *Brain Sci.* **2021**, *11*, 1012. https://doi.org/10.3390/brainsci11081012

Academic Editor: Corrado Romano

Received: 15 July 2021
Accepted: 27 July 2021
Published: 30 July 2021

Publisher's Note: MDPI stays neutral with regard to jurisdictional claims in published maps and institutional affiliations.

1. Introduction

There are an estimated 125,461 adults with Down syndrome (DS) living in the United States [1,2]. DS is associated with co-occurring medical conditions and variable intellectual disability (ID). Individuals with DS differ in the extent to which they can complete activities of daily living; and independence, the ability to complete tasks of self-care, varies in individuals with DS [3–5]. A variety of factors can contribute to function, such as: cognition, health, and social factors amongst others [5]. Survey has shown that those with more current health issues were significantly less likely to be independent and social; current health issues impact communication skills [5]. Importantly, communication skills vary in individuals with DS, including studies showing: 50% of individuals with DS speaking well by age 11 years, 10 months [6], 15–45% of adults with DS using verbal communication with no difficulty [5,7], 39% of children with DS expressing with no help required [8], and 42–58% of adults understanding verbal communication [5,7]. Surveys describing one's ability to complete various activities of daily living show a spectrum of independence in DS [5–7,9].

In describing the natural history of independence in individuals with DS, two studies of validated instruments emerge. One example is the Functional Independence Measure for Children (WeeFIM) questionnaire, which was used for children with DS and showed highest scores in mobility domain, and lowest in cognition domain [8]. A second example of an instrument related to independence is the Adaptive Behaviour Assessment System-II Adult (ABAS-II Adult) completed by parents and caregivers of adults with DS [10]. They found an association between increased age and lower adaptive behavior, suggesting that adults with DS may benefit from additional support in terms of their social and conceptual abilities as they age [10]. Beyond these two relevant instruments, we did not identify existing instruments which directly measure independence and are validated in adults with DS. Although studies suggest that independence and function decrease with age in adults with DS [7,9], interventions can be useful: speech training leads to increased autonomy and communication [11], and medical home access increased the odds of transition preparation and taking responsibility for health care [12].

Though the studies of independence in adulthood specific to DS are limited, lessons can be learned from the ID research literature. Natural history studies of those with ID have shown: the proportion of individuals with autism spectrum disorder (ASD) who are able to acquire a driver's license [13] and a link between autonomy to better health and health-related quality of life (HRQOL) [14]. Some features which predict factors related to independence include: physical fitness tests (manual dexterity, balance, comfortable and fast gait speed, muscular endurance, and cardiorespiratory fitness) and changes in activities of daily living (ADLs), predictive for a decline in ability in ID [15], and a poor social network was associated with worse health outcomes in older adults in the general population [16]. A few interventions to improve independence have been studied which include aids such as videos or digital technology: video prompting improves grocery shopping in ID [17], eating aids improve independent eating in ID [18], staff trained to teach those with ID to promote self-management [19], video self-modeling improves independence [20], and the use of tech and remote support services improve independence [21].

There is need for more study of independence in individuals with DS to describe the current level of independence, as studies identified rely on data collected nearly ten years ago or longer [5–7]. There is a need to better understand factors, such as communication, which contribute to independence in DS. Specifically, we began this descriptive study to understand the current skills in our clinic population of individuals with DS, with the ultimate goal of using this information to (1) gain awareness and understanding on the level of independence in our clinical cohort, (2) to identify targets for future quality improvement work to improve independence, and (3) to guide future research efforts which rely on use of communication. Knowing what skills are attained could have implications on research related to surveys, interviews, and instrument development, as well as research on broader topics such as the interplay between independence level and health status.

2. Materials and Methods

The Mass General Hospital Down Syndrome Program (MGH DSP), is a multidisciplinary, DS specialty clinic that provides comprehensive care to 550 unique patients annually. MGH DSP offers care for infants, children, adolescents, and adults with DS through distinct clinics: Infant and Toddler Clinic (ages birth–5 years), Child Clinic (ages 5–13 years), Adolescent and Young Adult Clinic (ages 13–21 years), and Adult Clinic (ages 21 years and older). MGH DSP also offers prenatal consultation for expectant parents. The multidisciplinary team of physicians, social workers, nutritionists, and many others care for a large volume of patients with DS. This study was part of a series of ongoing projects within the program to improve the quality of our work over time.

At the MGH DSP, caregivers are asked to complete an electronic clinic intake form in advance of their loved one's visit. The electronic intake form is shared through email, and is completed in a password-protected, Health Insurance Portability and Accountability Act (HIPAA)-secured platform which is housed through the MGH Laboratory of Computer Science; responses are stored securely on the MGH server. This electronic intake form was developed by the MGH DSP and collects a variety of pre-visit medical and lifestyle information electronically. This information is reviewed prior to the visit. We conducted a retrospective review of these electronic intake forms to evaluate patients' activities of daily living and independence; fields reviewed included: demographics (age, sex, race, co-occurring medical conditions), school/vocational setting, living situation, communication style, daily living skills, and independence skills.

Inclusion criteria: a patient at the MGH DSP from 2014–2020, with a completed electronic intake form. Patients without an electronic intake form were not included. While the MGH DSP sees patients across the country and internationally, the electronic clinic intake form was only available in English; therefore, parents with a primary language other than English do not complete the clinic intake form. Parents of returning patients age 5 and under who are following up every 6 months are not asked to complete an additional intake form at each visit because they are seen more often than older patients. Families who were seen at MGH DSP prenatally or who gave birth at MGH and had an inpatient consult are also not asked to complete an intake form. The MGH DSP does offer a mailed paper intake for those who request it, but this data was not included in the analysis.

Descriptive statistics included: mean, standard deviation, frequencies. Data was analyzed by the patient's age, with descriptive statistics for each subgroup reported. Data is available in de-identified, aggregate format from the author at reasonable request. This study was approved by the MassGeneral Brigham (formerly Partners) Institutional Review Board.

3. Results

From 2014–2020, 350 pediatric and 196 adult patients had a first visit to the MGH DSP and completed an intake form. Of these, 521 of the completed intake forms were completed by the caregiver of the individual with DS, and 25 of the intake forms were marked as self-completed by the individual with DS. Demographic details showed age ranging from 0–62 years and 46% of the patients were female. The average age of the overall cohort was 22 years, of which the average age of those completing the pediatric intake and adult intake were 10 and 35 years, respectively. The majority of the cohort self-identified as white (88%), and not Hispanic or Latino (83.3%). In the total cohort, the most common co-occurring medical condition reported was heart disease (38.6%). In the pediatric group alone, hypotonia was reported most frequently (40.8%). For adults only, heart disease remained most common, followed by thyroid disease (44.9%). (Table 1).

Table 1. Demographic details of the 546 patients with Down syndrome (DS) in the Massachusetts General Hospital Down Syndrome Program (MGH DSP) with completed electronic intake forms.

	Total Cohort (*n* (%))	Pediatric Intakes (*n* (%))	Adult Intakes (*n* (%))
Male	295 (54%)	185 (53%)	110 (56%)
Race:			
White	482 (88%)	297 (85%)	185 (94%)
Black or African American	19 (4%)	14 (4%)	5 (3%)
American Indian	2 (<1%)	2 (1%)	0
Asian	52 (10%)	48 (14%)	4 (2%)
Hawaiian	1 (<1%)	1 (0.3%)	0
Other	18 (3%)	17 (5%)	1 (1%)
Ethnicity:			
Hispanic or Latino	91 (17%)	87 (25%)	4 (2%)
Co-occurring medical conditions:			
Heart disease	211 (39%)	117 (33%)	94 (48%)
Seizures	32 (6%)	13 (4%)	19 (10%)
Autism	42 (8%)	29 (8%)	13 (7%)
Dementia	16 (3%)	0	16 (8%)
Cognitive Decline	49 (9%)	12 (3%)	37 (19%)
Expressive Language Delay	125 (23%)	71 (20%)	54 (28%)
Hypotonia	190 (35%)	143 (41%)	47 (24%)
Thyroid disease	179 (33%)	91 (26%)	88 (45%)
Depression	51 (9%)	12 (3%)	39 (20%)
Anxiety	104 (19%)	50 (14%)	54 (28%)

Table 1. *Cont.*

	Total Cohort (*n* (%))	Pediatric Intakes (*n* (%))	Adult Intakes (*n* (%))
Obsessive Compulsive Disorder (OCD)	65 (12%)	23 (7%)	42 (21%)
Attention-Deficit/Hyperactivity Disorder (ADHD)	34 (6%)	28 (8%)	6 (3%)
Post-traumatic stress disorder	9 (2%)	4 (1%)	5 (3%)
	Mean	Mean	Mean
Age (years)	22.3	10.2	35.4

Daily characteristics of the sample included: pediatric patients were most often enrolled in school, and in an inclusion setting, while adult patients were most often participating in a day program, living with family, and wanted to continue living with family in the future (Table 2). Most (87%) of adults with DS communicated verbally, though fewer used written communication (17%). Assistive devices were used by both pediatric (34%) and adults (30%) with DS. Adults with DS used a smartphone (52%), read (58%), and wrote (64%). Activities of daily living which were completed by the adult with DS on his/her own ranged: 4% could cook independently, 46% bathed independently, 48% showered independently, 58% brushed teeth independently, 71% toileted independently, 74% dressed independently, and 89% fed independently (Table 2).

Table 2. Characteristics of 350 pediatric and 196 adult patients with Down syndrome (DS) in the Massachusetts General Hospital Down Syndrome Program (MGH DSP).

PEDIATRIC (0–18 Years)		*n*	%
Currently enrolled inschool		265	76
If so, placement:	Inclusion	92	35
	Partial inclusion	76	29
	Separate	73	28
	Collab	21	8
	Cotaught	9	3
	Homeschool	7	3
If so, setting:	Private	25	9
Therapies	Speech therapy at school	236	67
	Speech therapy at home	114	33
	Physical therapy at school	172	49
	Physical therapy at home	83	24
	Occupational therapy at school	201	57
	Occupational therapy at home	89	25
	Behavioral therapy at school	59	17
	Behavioral therapy at home	24	7
	Vocational therapy at school	54	15
	Vocational therapy at home	11	3
	Job coaching at school	32	9
	Job coaching at home	8	2
Has IEP		252	72
Satisfied with IEP		199	79
Uses Assistive Device		119	34
ADULT (19 years and older)			
Which best describes the current living situation	Living on his/her own	6	3
	Living on his/her own with a Supported Living Coordinator	11	6
	Living with a family member	140	71
	Living with a roommate	3	2
	Living in a group home	26	13
	Other	10	5

Table 2. *Cont.*

PEDIATRIC (0–18 Years)		*n*	%
	Live on own	16	8
	Live on own with a Supported Living Coordinator	30	15
Where he/she would like to live in the future (check all that apply):	Live with a family member	98	50
	Live with a roommate	27	14
	Live in a group home	52	27
	Other	14	7
He/she uses an assistive device	Yes	58	30
	Currently enrolled in school	19	10
Vocation	Currently enrolled in day program	127	65
	Currently employed	96	49
	Verbal	170	87
	Gesture	72	37
	Sign	31	16
He/she communicates via (check all that apply):	Pictures	22	11
	Written	34	17
	Device	11	6
	Other	17	9
Does he/she use a smartphone or other mobile device?	Yes	101	52
	Feeding	175	89
	Dressing	146	74
	Bathing	90	46
Is he/she able to do on his or her own?	Toileting	139	71
	Showering	94	48
	Brushing teeth	114	58
	Cooking	7	4
Is he/she able to read?	Yes	114	58
Is he/she able to write?	Yes	126	64

Independence skills for adolescents and adults with DS were ranked by importance, and those of greatest importance included: learning about healthy foods (35%), preparing meals (34%), providing personal information when needed (35%), and describing symptoms to a doctor (35%; Table 3). Skills which were most often reported as attained included: dressing self (72%), getting 7 to 8 h of sleep (58%), using a public restroom (56%), and swallowing whole pills (53%). Some skills were most often reported as not important: learning how to refill my prescriptions on my own (69%), learning what each medicine is for (61%), finding my medication list (60%). Many skills show a range in responses across the response options in the electronic intake form filled out before their clinic visit.

Table 3. Independence skills of 337 patients with Down syndrome age 13+ years in the MGH DSP.

	NI [+]	TL [‡]	RIN [^]	SA [§]	Attainment Was Described as:
I want to learn about the differences between healthy and unhealthy foods.	83	48	119	87	I know which foods are healthy and I try to pick the healthy foods for my meals.
I want to learn how to provide my personal information (name, emergency contact person) when needed (for example, if I get lost and a police officer asks for my name).	66	42	119	110	I am able to provide my personal information when needed without any help.
I want to learn how to describe how I am feeling to my doctor (for example, "I feel pain", "I'm having a hard time breathing", or "I'm coughing")	89	65	118	65	I am always able to tell the doctor how I feel.
I want to be able to prepare my own meals.	93	96	116	32	I am able to prepare my own meals.

Table 3. *Cont.*

	NI [+]	TL [‡]	RIN [^]	SA [§]	Attainment Was Described as:
I want to exercise regularly.	65	41	112	119	I exercise regularly.
I want to learn how to tell the difference between a stranger and a friend.	72	38	106	121	I know how to tell the difference between a stranger and a friend.
I want to be able to bathe/shower myself.	51	30	101	155	I am able to bathe or shower without any help.
I want to understand sexual boundaries and privacy.	100	63	96	78	I believe I have a good understanding of sexual boundaries and privacy.
I want to have a plan for what I will do after finishing high school (e.g., more school, work, career goals).	113	47	93	84	I have figured out what I am doing after high school.
I want to learn how to brush my teeth on my own.	48	26	90	173	I always brush my teeth without help.
I want to learn how to call 911 if there is an emergency.	73	54	83	127	I know how to call 911 in an emergency.
I want to be able to do my laundry.	116	96	72	53	I am able to do my own laundry.
I want to learn how to do household chores.	64	46	70	157	I always help with household chores.
I want to sleep 7 to 8 h per night.	65	13	65	194	I always get 7 to 8 h of sleep at night.
I would like to name at least two adults I can ask for help when I need it.	71	30	62	174	I am able to name two adults I can go to for help.
I want to be able to use a public restroom on my own.	63	30	55	189	I am comfortable using a public restroom alone.
I want to learn how to ask my doctor questions.	130	96	54	57	I am able to ask my doctor questions on my own.
I want to be able to dress myself.	34	16	43	244	I get dressed myself.
I want to learn how to use public transportation on my own.	187	95	38	17	I am able to use public transportation when I am on my own.
I want to be able to find my medication list.	201	71	36	29	I always know where to find my medication list.
I want to learn what each of my medicines is for (for example, "I take Synthroid for my thyroid").	205	55	33	44	I know why I take all of my medications.
I want to learn about the risks of alcohol, drugs and tobacco use.	193	46	27	71	I know how alcohol, drugs and tobacco could affect my body.
I want to learn where to find my doctor's phone number.	192	87	26	32	I always know where to find my doctor's phone number.
I want to learn how to swallow whole pills.	97	38	24	178	I can swallow whole pills.
I want to take my medications every day on my own.	169	59	23	86	I can take my medications with little or no help.
I want to learn how to manage my period. (for females)	44	18	22	74	I can usually manage my period without any help.
I want to learn how to find my insurance card.	178	98	19	42	I always know where my insurance card is.
I want to learn how to refill my prescriptions on my own.	233	73	17	14	I can refill my prescriptions with little or no help.

[+] Note important. [‡] Not important now to me now, but I want to try later. [^] This is really important to me now. [§] Skill attained. Note: Instructions: this section should be completed by or with the patient with Down syndrome. If a particular question does not apply to the patient, choose 'Not important'. Gray highlighting = most frequent response to each item.

Among those skills ranked as really important now, but not yet attained, additional questions were asked about level of ability. Those with higher proportion showing progress to attainment included: providing personal information (71%), learning about healthy foods (71%), learning to do household chores (90%), while those with smaller proportion showing progress to attainment included: asking my doctor questions (70%), finding medication list (73%), refilling prescriptions (88%), and swallowing whole pills (70%; Table 4).

Table 4. Skill level if "This is really important to me now." This corresponds to Table 3 about individuals with Down syndrome 13+ years in MGH DSP.

Topic	Response	N
Find my doctor's phone number.	I need a lot of help finding my doctor's phone number.	14
	I need a little help finding my doctor's phone number.	9
Ask my doctor questions.	My parent or guardian usually asks questions for me.	38
	My parent or guardian usually reminds me to ask some questions.	16
Find my medication list.	I rely on someone else to keep the list and know my medications.	24
	I know where to find my medication list, but sometimes I forget and need some help.	9
Refill my prescriptions on my own.	I rely on my parent or guardian to refill all my medications.	15
	I sometimes help my parent or guardian call in and pick up refills of my medications.	2
Swallow whole pills.	I am unable to swallow pills and usually need to have them crushed.	17
	I am learning how to swallow a pill.	6
Tell the difference between a stranger and a friend.	I am not able to tell the difference between a stranger and a friend.	58
	I can tell the difference between a stranger and friend with help.	48
Sleep 7 to 8 h per night.	I do not usually get enough sleep at night.	41
	I am learning the importance of getting enough sleep at night.	20
Use public transportation on my own.	I do not use public transportation when I am alone.	26
	I am learning how to use public transportation without a parent or guardian.	12
Learn how to manage my period. (for females)	I need help using pads or tampons when I have my period.	11
	I am learning how to manage my period without help.	10
Brush my teeth on my own.	I need help brushing my teeth.	50
	I am learning to brush my teeth without help.	39
Use a public restroom on my own.	I need help using a public restroom.	31
	I am learning how to use a public restroom without help.	23
Find my insurance card.	I do not know where my insurance card is.	11
	I am learning where my insurance card is kept.	7
Describe how I am feeling to my doctor.	My parent or guardian usually explains how I'm feeling to the doctor.	51
	I try to describe how I am feeling and if I have trouble, my parents help me.	65
Learn what each of my medicines is for.	I do not know what any of my medications are for.	11
	I know why I am taking some of my medications.	22
Take my medications every day on my own.	I am not really sure when or how to take my medications.	8
	I am learning when to take my medications and which medications to take in the morning, noon or night (for example, I have a pill organizer that I am learning to use).	15
Provide my personal information (name, emergency contact person) when needed.	I am not able to provide any information (either verbally or nonverbally).	34
	If asked, I am able to provide some of my information or pull out an identification card that has all my information.	83
Learn about the differences between healthy and unhealthy foods.	I have trouble telling the difference between healthy and unhealthy foods.	34
	I am learning about what foods are healthy and which are not as healthy for me.	83
Learn about the risks of alcohol, drugs and tobacco use.	I do not know how alcohol, drugs and tobacco can affect me.	13
	I am learning about how alcohol, drugs and tobacco can affect me.	14
Call 911 if there is an emergency.	I do not know how to call 911 in an emergency.	35
	I am learning how to get help and call 911 in an emergency.	45
Exercise regularly.	I do not exercise regularly.	44
	I am trying to exercise regularly.	66
Learn how to do household chores.	I do not help with household chores.	7
	I am learning to help with some chores.	63
Understand sexual boundaries and privacy.	I struggle with sexual boundaries and privacy.	27
	I am learning about sexual boundaries and privacy.	65
Dress myself.	I need a lot of help getting dressed.	13
	I need a little help getting dressed.	30
Bathe/shower myself.	I need a lot of help taking a bath or with showering.	42
	I need a little help taking a bath or with showering.	58
Prepare my own meals.	Someone else prepares my meals for me.	48
	I am learning to prepare my own meals.	68
Do my laundry.	Someone else does my laundry.	22
	I am able to help with the laundry (for example, sorting or folding laundry).	49
Plan for what I will do after finishing high school.	I do not know what I am doing after high school.	29
	I am working on figuring out what to do after high school.	63
Name at least two adults I can ask for help.	I cannot name two adults I can go to for help.	19
	I am working on identifying two adults I can go to for help.	41

Note: grey highlight = more responses of minimal attainment; yellow highlight = more responses of moderate attainment.

When analyzing the percent of individuals of a certain age who had attained a given skill, we saw that individuals with DS gained skills into adulthood (Table 5). Through the lifespan range, half or more of patients could get dressed on their own. Many skills were able to be completed by half or more of the individuals age 40–49 years. Some skills were less often completed through the lifespan, such as refilling prescriptions and using public transportation.

Table 5. Independence skills attained by age group of 337 patients with Down syndrome age 13+ years in the MGH DSP.

| | Age Group (Years) | | | | | | | |
| | 13–17 | 18–22 | 23–29 | 30–39 | 40–49 | 50–59 | 60+ | Total |
Attainment Was Described as:	$n = 83$	$n = 91$	$n = 59$	$n = 45$	$n = 24$	$n = 31$	$n = 4$	
			Skill Attained in … n (%)					
I get dressed myself.	60 (72)	55 (60)	49 (83)	40 (89)	19 (79)	17 (55)	2 (50)	244
I always get 7 to 8 h of sleep at night.	49 (59)	49 (54)	35 (59)	30 (67)	14 (58)	14 (45)	2 (50)	194
I am comfortable using a public restroom alone.	42 (51)	40 (44)	39 (66)	35 (78)	17 (71)	13 (42)	1 (25)	189
I can swallow whole pills.	42 (51)	39 (43)	37 (63)	25 (56)	14 (58)	16 (52)	2 (50)	178
I always brush my teeth without help.	42 (51)	36 (40)	31 (53)	30 (67)	17 (71)	15 (48)	1 (25)	173
I am able to bathe or shower without any help.	35 (42)	36 (40)	28 (47)	30 (67)	14 (58)	9 (29)	1 (25)	155
I am able to name two adults I can go to for help.	34 (41)	41 (45)	33 (56)	35 (78)	18 (75)	13 (42)	0 (0)	174
I always help with household chores.	31 (37)	39 (43)	28 (47)	28 (62)	12 (50)	16 (52)	2 (50)	157
I exercise regularly.	28 (34)	34 (37)	22 (37)	21 (47)	7 (29)	7 (23)	0 (0)	119
I know how to call 911 in an emergency.	25 (30)	29 (32)	20 (34)	29 (64)	14 (58)	9 (29)	0 (0)	127
I know how to tell the difference between a stranger and a friend.	21 (25)	28 (31)	21 (36)	22 (49)	15 (63)	13 (42)	1 (25)	121
I know which foods are healthy and I try to pick the healthy foods for my meals.	19 (23)	16 (18)	13 (22)	22 (49)	11 (46)	6 (19)	0 (0)	87
I am able to provide my personal information when needed without any help.	18 (22)	23 (25)	26 (44)	21 (47)	13 (54)	9 (29)	0 (0)	110
I can take my medications with little or no help.	15 (18)	23 (25)	16 (27)	14 (31)	11 (46)	5 (16)	0 (0)	86
I know how alcohol, drugs and tobacco could affect my body.	14 (17)	13 (14)	13 (22)	19 (42)	7 (29)	5 (16)	0 (0)	71
I am always able to tell the doctor how I feel.	13 (16)	13 (14)	12 (20)	11 (24)	7 (29)	8 (26)	0 (0)	65
I am able to ask my doctor questions on my own.	11 (13)	16 (18)	8 (14)	12 (27)	5 (21)	5 (16)	0 (0)	57
I know why I take all of my medications.	11 (13)	10 (11)	10 (17)	8 (18)	3 (13)	2 (6)	0 (0)	44
I always know where to find my doctor's phone number.	7 (8)	6 (7)	6 (10)	8 (18)	2 (8)	3 (10)	0 (0)	32
I believe I have a good understanding of sexual boundaries and privacy.	6 (7)	15 (16)	16 (27)	26 (58)	8 (33)	7 (23)	0 (0)	78
I have figured out what I am doing after high school.	3 (4)	16 (18)	25 (42)	26 (58)	9 (38)	4 (13)	0 (0)	84
I always know where to find my medication list.	3 (4)	5 (5)	4 (7)	9 (20)	6 (25)	2 (6)	0 (0)	29
I am able to do my own laundry.	3 (4)	9 (10)	9 (15)	16 (36)	9 (38)	6 (19)	0 (0)	53
I am able to prepare my own meals.	2 (2)	6 (7)	3 (5)	8 (18)	7 (29)	6 (19)	0 (0)	32
I always know where my insurance card is.	1 (1)	7 (8)	8 (14)	13 (29)	3 (13)	8 (26)	1 (25)	42
I can refill my prescriptions with little or no help.	1 (1)	3 (3)	1 (2)	4 (9)	4 (17)	1 (3)	0 (0)	14
I am able to use public transportation when I am on my own.	1 (1)	1 (1)	3 (5)	8 (18)	4 (17)	0 (0)	0 (0)	17
I can usually manage my period without any help.	24	14	18	10	6	1	0	74

Note: this section should be completed by or with the patient with Down syndrome. If a particular question does not apply to the patient, choose 'Not important'. Green ≥ 50% attained; Yellow = 25–49% attained; Red = 10–24% attained; Gray < 10% attained.

4. Discussion

Through retrospective review of the electronic clinic intake forms of 546 patients with DS in the MGH DSP, we found:

- There was great variability in activities of daily living completed by an adult with DS on their own, and most of our adults with DS used verbal communication.
- Dressing self, sleeping 7 to 8 h a night, using a public restroom on their own, and swallowing whole pills were the independence skills most often attained.
- Skills of highest importance to our patients were learning about healthy foods, preparing their own meals, communicating personal information, and describing symptoms to a doctor.

Adults with DS exhibited a great range in the type and number of activities of daily living that they were able to complete. Fewer adults with DS cooked independently, while some bathed and showered independently, and the majority were able to brush their teeth, use the restroom, and eat independently. A similar study of adults with DS separated meal preparation abilities into two categories, preparing simple meals and cooking meals; 18% said they had "a lot of difficulty" preparing simple meals like sandwiches or cereal, compared to 52.2% that said they had "a lot of difficulty" with cooking [5]. While we did not make this distinction in our electronic intake form, both rates of independence in Matthews et al. are greater than the findings in our cohort in which 4% of adults with DS were able to cook on their own. In our cohort, 46% of adults with DS bathed on their own and 48% showered on their own. A study of children with DS which looked at the level of supervision needed for bathing found similar results, with 48% requiring no help [8]. We found that 71% of adults with DS in our cohort used the restroom independently, which aligns to a previous study in which 76.4% of adults with DS used the toilet independently [5]. Lastly, we found that 89% of our adult cohort ate independently; a previous study found that 88.8% of their adult cohort could eat independently [5]. This description of activities of daily living helps to quantify the level of support needed on a daily basis for adults with DS in our cohort, which could be useful for future studies aiming at improving aspects of independence.

The preferred form of communication varies in DS; many (87%) of our adults with DS used verbal communication. Previous study found 92.3% of individuals with DS age 14–62 used verbal communication to some extent, though their results include granularity on the level of difficulty ranging from 15% who used verbal communication with no difficulty, to 18% who used verbal communication with great difficulty [7]. Additionally, in our cohort, 58% could read and 64% could write; these rates are higher than published (8.1–52.1% and 13.5–52.1%, respectively) [6,7].

The independence skills most often attained by our patients with DS age 13 and up were dressing, sleeping 7 to 8 h a night, using a public restroom on their own, and swallowing whole pills. In our cohort, 72% of patients with DS over the age of 13 dressed themselves; this rate was less than that which De Graaf et al. found, which was that 83.9% of adults studied with DS over the age of 20 were able to dress with no additional help [6]. In our cohort, 58% of our patients with DS age 13 years and older slept 7 to 8 h a night. Although studies evaluating sleep have measured sleep duration using methods such as actigraphy watches and parent report, the other studies we identified which evaluate aspects of independence have not collected information on sleeping abilities, despite the high prevalence of sleep apnea within the population [22]. Importantly, although a previous study found that 35% of adults with DS were able to take medications independently, in our cohort, more than half (53%) could swallow pills whole on their own, which has important implications for feasibility of medication administration in future clinical trials and medication adherence [5]. In addition, we found that 56% of patients in our cohort can use the public restroom on their own; this has not been evaluated in previous studies on independence, but has important implications on integrating adults with DS into the community, such as the ability to take outings in the community alone, to navigate public situations independently, and to go into settings which might require the use of a public restroom.

When planning for the future, the focus will likely be on unattained but important skills. In our cohort, the skills that were reported as most important and not yet attained were: learning about healthy foods, preparing their own meals, communicating personal information, and describing symptoms to a doctor. Those of least importance included learning to refill prescriptions, knowing what each medication is for, and finding their medication list. While other studies have looked at the level of attainment of some of these skills, none have reported on which were most important and least important for families to achieve with their loved one with DS. Behavioral support and interventions would likely best be focused on some of the unattained but important skills rather than those of lesser importance. For example, given the interest in healthy foods and meal preparation, it might be important to give additional resources to support nutritionists, feeding therapists, and related resources, like adapted cookbooks or electronic mealtime supports [23]. Resources are available online to promote independence skills [24]. Given the interest in communicating personal information, it might be helpful to create a tool for individuals with DS to practice this skill and could even be combined with the skill of describing symptoms to a doctor. One online digital healthcare tool developed by the MGH Down Syndrome Program and MGH Laboratory of Computer Science, Down Syndrome Clinic to You (DSC2U), provides families with a loved one with DS 13 years or older with a life skills checklist to identify social stories and other resources to practice and improve the life skills families select as most important right now [25]. For example, if a caregiver marks that their teenager or adult with DS would like to work on describing how they feel to their doctor, one skill we found is very important to families for independence, DSC2U will suggest specific resources to help work on this specific independence goal. If they select that they would like to work on preparing meals on their own, DSC2U will link to a cookbook designed specifically for people with DS. DSC2U can bring tailored independence resources directly to families to improve on a variety of life skills [25].

It is also interesting to note that the skills rated as unattained and least important were most often related to medical encounters. Instead of discounting these skills as not important for families, these could be growing points for physicians to support families in realizing areas their loved ones can have independence in medical settings. While a skill like refilling a prescription may feel unattainable or unnecessary at the time due to familial support, that is one area that the person with DS could exert independence with encouragement from family and providers. Having this information on the areas of greatest and least importance seems to hold opportunities for future research and intervention.

Notably, the percent of individuals of a given age who had attained given skills was high in our subgroups in their 30s and 40s. Adults in their 30s and 40s more often attained skills (in half or more of the cohort) than adolescents age 13–17 years and young adults age 18–22 years. While our cohort of adults in their 40s was still able to carry out many skills of independence, attainment seems to fall off for many of skills in the 50–59 and 60+ age groups. Differences between individuals and age groups may account for some of this variation in attainment of skill, and future studies should follow individuals longitudinally to determine if individuals can indeed gain independence skills through adulthood. However, our analysis by age subgroup provides useful evidence that adults with DS can continue to gain skills after high school and into adulthood.

One purpose of our study was to develop a baseline for future quality improvement goals for our program and to identify targets for future quality improvement work to improve independence. Our study is limited by our data collection method and the aspects of our single cohort. The electronic clinic intake form was mostly completed by parents or caregivers, but some marked that the form was self-completed by the individual with DS. So, all information collected was caregiver-reported or self-reported and not confirmed through a validated instrument or clinical notes. We cannot verify the extent of assistance, if any, with which the intake form was self-completed, therefore we are unable to distinguish whether caregivers read it to the person with DS, explained questions, or if the individual with DS completely independently completed the intake process. Although the MGH DSP sees patients across the country and internationally, non-English speakers were not included since the electronic intake form was only available in English at the time of the study. In the case of technology limitations or difficulties, caregivers were presented with the option of a mailed paper version of the form. These non-electronic, paper versions were not included in our data analysis. Additionally, our study did not distinguish between different types of heart disease; 39% had heart disease within our cohort, but this was not specifically congenital heart disease. This study reflects just one cohort of patients at our program and may not generalize to other clinics, or all individuals with DS. Future studies could expand on this review to include additional DS specialty clinics, or to track development of independence skills in our cohort over time.

In the future, it would be useful to collect self-reported data from individuals with DS of what skills are most important and most meaningful for independence, rather than caregiver-reported data. These skills can then guide future surveys on independence, future instruments to measure health, and future modifications to care. For example, the intake form did not include many questions about social activities which were asked in other studies of independence, such as working, volunteering, hanging out with friends, or playing games. If this is viewed as important to patients and their families, it could be added to better capture aspects of life which are meaningful to independence. The study provides information on the skills which are viewed as the most meaningful for independence and lays a framework for beginning to develop a measure of independence in DS. If an instrument was developed to measure independence reliably and validly, this could have important implications on future research and interventions. In comparing our results to published studies, we found that it was not easy to combine data with other independence studies given the varied methods of data collection, varied wording on surveys used, and varied approaches to reporting results as the percentage attaining a skill or the age at which a skill was attained. Developing a standard method of collection for independence skills and activities of daily living would allow researchers to expand study to include larger cohorts such as a national sample through DS-Connect [26]. Finally, with this knowledge, studies could be done to focus on how best to modify those factors which are most important for independence and daily living. Identifying effective interventions and ways to support families in building these skills could guide DS specialty clinics, researchers studying DS, and parent resource groups and advocacy organizations dedicated to individuals with DS.

5. Conclusions

Independence skills for patients with DS are varied; those used during a medical appointment could be improved. Skills of greatest importance should be the focus of future research and intervention.

Author Contributions: Conceptualization, S.L.S.; methodology, S.L.S.; software, A.T.; validation, A.T.; formal analysis, K.K., K.H. and S.L.S.; investigation, K.K., K.H. and S.L.S.; resources, S.L.S.; data curation, K.H.; writing—original draft preparation, K.K.; writing—review and editing, K.K., K.H., A.T. and S.L.S.; visualization, S.L.S.; supervision, A.T.; project administration, A.T.; funding acquisition, S.L.S. All authors have read and agreed to the published version of the manuscript.

Funding: This research was funded by NIH NICHD Eunice Kennedy Shriver, grant number K23HD100568.

Institutional Review Board Statement: The study was conducted according to the guidelines of the Declaration of Helsinki and approved by the Institutional Review Board of Approved by IRB of Mass General Brigham, formerly Partners Healthcare (protocol code 2020P003890 and 15 December 2020).

Informed Consent Statement: Patient consent was waived because the consent would be the only document linking patients to the research and the purpose of this is developing a baseline for future quality improvement goals.

Data Availability Statement: The data presented in this study are available on request from the corresponding author. The data are not publicly available because we did not notify survey respondents of data sharing at time of completing the survey, therefore, they did not explicitly consent to data sharing.

Acknowledgments: Appreciation is given to the Laboratory for Computer Sciences for creating the electronic intake form, and to Brian Skotko, the Director of the MGH Down Syndrome Program, for his leadership and foresight to implement use of this intake form.

Conflicts of Interest: The authors declare no conflict of interest. The funders had no role in the design of the study; in the collection, analyses, or interpretation of data; in the writing of the manuscript, or in the decision to publish the results.

References

1. Santoro, S.L.; Campbell, A.; Balasubramanian, A.; Haugen, K.; Schafer, K.; Mobley, W. Specialty clinics for adults with Down syndrome: A clinic survey. *Am. J. Med. Genet. A* **2021**. [CrossRef] [PubMed]
2. De Graaf, G.; Buckley, F.; Skotko, B.G. Estimation of the number of people with Down syndrome in the United States. *Genet. Med.* **2017**, *19*, 439–447. [CrossRef]
3. Carr, J. Long-term-outcome for people with Down's syndrome. *J. Child. Psychol. Psychiatry* **1994**, *35*, 425–439. [CrossRef] [PubMed]
4. Hawkins, B.A.; Eklund, S.J.; James, D.R.; Foose, A.K. Adaptive Behavior and Cognitive Function of Adults with down Syndrome: Modeling Change with Age. *Ment. Retard.* **2003**, *41*, 7–28. [CrossRef]
5. Matthews, T.J.; Allain, D.C.; Matthews, A.L.; Mitchell, A.; Santoro, S.L.; Cohen, L. An assessment of health, social, communication, and daily living skills of adults with Down syndrome. *Am. J. Med Genet. Part A* **2018**, *176*, 1389–1397. [CrossRef] [PubMed]
6. De Graaf, G.; Levine, S.P.; Goldstein, R.; Skotko, B.G. Parents' perceptions of functional abilities in people with Down syndrome. *Am. J. Med. Genet. Part A* **2018**. [CrossRef]
7. Bertoli, M.; Biasini, G.; Calignano, M.T.; Celani, G.; De Grossi, G.; Digilio, M.C.; Fermariello, C.C.; Loffredo, G.; Luchino, F.; Marchese, A.; et al. Needs and challenges of daily life for people with Down syndrome residing in the city of Rome, Italy. *J. Intellect. Disabil. Res.* **2011**, *55*, 801–820. [CrossRef]
8. Lin, H.-Y.; Chuang, C.-K.; Chen, Y.-J.; Tu, R.-Y.; Chen, M.-R.; Niu, D.-M.; Lin, S.-P. Functional independence of Taiwanese children with Down syndrome. *Dev. Med. Child. Neurol* **2016**, *58*, 502–507. [CrossRef]
9. Esbensen, A.J.; Seltzer, M.M.; Krauss, M.W. Stability and change in health, functional abilities, and behavior problems among adults with and without Down syndrome. *Am. J. Ment. Retard.* **2008**, *113*, 263. [CrossRef]
10. Makary, A.T.; Testa, R.; Tonge, B.J.; Einfeld, S.L.; Mohr, C.; Gray, K.M. Association between adaptive behaviour and age in adults with Down syndrome without dementia: Examining the range and severity of adaptive behaviour problems. *J. Intellect. Disabil. Res.* **2015**, *59*, 689–702. [CrossRef]
11. Barbosa, T.M.M.F.; Lima, I.L.B.; Alves, G.Â.D.S.; Delgado, I.C. Contributions of speech-language therapy to the integration of individuals with Down syndrome in the workplace. *Codas* **2018**, *30*, e20160144. [CrossRef] [PubMed]
12. Nugent, J.; Gorman, G.; Erdie-Lalena, C.R. Disparities in access to healthcare transition services for adolescents with Down syndrome. *J. Pediatr.* **2018**, *197*, 214–220. [CrossRef] [PubMed]
13. Curry, A.E.; Yerys, B.E.; Huang, P.; Metzger, K.B. Longitudinal study of driver licensing rates among adolescents and young adults with autism spectrum disorder. *Autism* **2018**, *22*, 479–488. [CrossRef]
14. Alonso-Sardón, M.; Iglesias-de-Sena, H.; Fernández-Martín, L.C.; Mirón-Canelo, J.A. Do health and social support and personal autonomy have an influence on the health-related quality of life of individuals with intellectual disability? *BMC Health Serv. Res.* **2019**, *19*, 63. [CrossRef] [PubMed]
15. Oppewal, A.; Hilgenkamp, T.I.M.; van Wijck, R.; Schoufour, J.D.; Evenhuis, H.M. Physical fitness is predictive for a decline in the ability to perform instrumental activities of daily living in older adults with intellectual disabilities: Results of the HA-ID Study. *Res. Dev. Disabil.* **2015**, *41–42*, 76–85. [CrossRef] [PubMed]
16. Sakurai, R.; Kawai, H.; Suzuki, H.; Kim, H.; Watanabe, Y.; Hirano, H.; Ihara, K.; Obuchi, S.; Fujiwara, Y. Poor social network, not living alone, is associated with incidence of adverse health outcomes in older adults. *J. Am. Med. Dir. Assoc.* **2019**, *20*, 1438–1443. [CrossRef] [PubMed]
17. Bouck, E.C.; Satsangi, R.; Bartlett, W. Supporting grocery shopping for students with intellectual disability: A preliminary study. *Disabil. Rehabil. Assist. Technol.* **2017**, *12*, 605–613. [CrossRef] [PubMed]
18. Dag, M.; Svanelöv, E.; Gustafsson, C. Experiences of using bestic, an eating aid for people with intellectual disabilities. *J. Intellect. Disabil.* **2017**, *21*, 87–98. [CrossRef]
19. Sandjojo, J.; Zedlitz, A.M.E.E.; Gebhardt, W.A.; Hoekman, J.; Dusseldorp, E.; den Haan, J.A.; Evers, A.W.M. Training staff to promote self-management in people with intellectual disabilities. *J. Appl. Res. Intellect. Disabil.* **2018**, *31*, 840–850. [CrossRef]
20. Allen, K.D.; Vatland, C.; Bowen, S.L.; Burke, R.V. An evaluation of parent-produced video self-modeling to improve independence in an adolescent with intellectual developmental disorder and an autism spectrum disorder: A controlled case study. *Behav. Modif.* **2015**, *39*, 542–556. [CrossRef]
21. Tassé, M.J.; Wagner, J.B.; Kim, M. Using technology and remote support services to promote independent living of adults with intellectual disability and related developmental disabilities. *J. Appl. Res. Intellect. Disabil.* **2020**, *33*, 640–647. [CrossRef] [PubMed]
22. Esbensen, A.J.; Hoffman, E.K.; Stansberry, E.; Shaffer, R. Convergent validity of actigraphy with polysomnography and parent reports when measuring sleep in children with Down syndrome. *J. Intellect. Disabil. Res.* **2018**, *62*, 281–291. [CrossRef] [PubMed]
23. Lazar, J. Co-design process of a smart phone app to help people with Down syndrome manage their nutritional habits. *J. Usability Stud.* **2018**, *13*, 73–93.
24. Patient Resources for the Down Syndrome Program. Available online: https://www.massgeneral.org/children/down-syndrome/patient-handouts (accessed on 1 July 2021).

25. Chung, J.; Donelan, K.; Macklin, E.A.; Schwartz, A.; Elsharkawi, I.; Torres, A.; Hsieh, Y.G.; Parker, H.; Lorenz, S.; Patsiogiannis, V.; et al. A randomized controlled trial of an online health tool about Down syndrome. *Genet. Med.* **2020**. [CrossRef]
26. Peprah, E.K.; Parisi, M.A.; Kaeser, L.; Bardhan, S.; Oster-Granite, M.; Maddox, Y.T. DS-connect: A promising tool to improve lives and engage down syndrome communities worldwide. *Glob. Heart* **2015**, *10*, 337–340. [CrossRef]

Article

Acute Regression in Down Syndrome

Benjamin Handen [1,*], Isabel Clare [2], Charles Laymon [1], Melissa Petersen [3], Shahid Zaman [2], Sid O'Bryant [3], Davneet Minhas [1], Dana Tudorascu [1], Stephanie Brown [2], and Bradley Christian [4] on behalf of the Alzheimer's Biomarker Consortium-Down Syndrome (ABC-DS)

1   Department of Psychiatry, University of Pittsburgh, Pittsburgh, PA 15260, USA; cml14@pitt.edu (C.L.); dam148@pitt.edu (D.M.); dlt30@pitt.edu (D.T.)
2   Department of Psychiatry, University of Cambridge, Cambridge CB2 1TN, UK; ichc2@medschl.cam.ac.uk (I.C.); shz10@medschl.cam.ac.uk (S.Z.); sb2403@medschl.cam.ac.uk (S.B.)
3   Department of Family Medicine, University of North Texas Health Science Center, Fort Worth, TX 76107, USA; Melissa.Petersen@unthsc.edu (M.P.); SidObryant@unthsc.edu (S.O.)
4   Departments of Medical Physics and Psychiatry, University of Wisconsin, Madison, WI 53706, USA; bchristian@wisc.edu
*   Correspondence: handenbl@upmc.edu; Tel.: +1-412-235-5445

Abstract: Background: Acute regression has been reported in some individuals with Down syndrome (DS), typically occurring between the teenage years and mid to late 20s. Characterized by sudden, and often unexplained, reductions in language skills, functional living skills and reduced psychomotor activity, some individuals have been incorrectly diagnosed with Alzheimer's disease (AD). Methods: This paper compares five individuals with DS who previously experienced acute regression with a matched group of 15 unaffected individuals with DS using a set of AD biomarkers. Results: While the sample was too small to conduct statistical analyses, findings suggest there are possible meaningful differences between the groups on proteomics biomarkers (e.g., NfL, total tau). Hippocampal, caudate and putamen volumes were slightly larger in the regression group, the opposite of what was hypothesized. A slightly lower amyloid load was found on the PET scans for the regression group, but no differences were noted on tau PET. Conclusions: Some proteomics biomarker findings suggest that individuals with DS who experience acute regression may be at increased risk for AD at an earlier age in comparison to unaffected adults with DS. However, due to the age of the group (mean 38 years), it may be too early to observe meaningful group differences on image-based biomarkers.

Keywords: down syndrome; regression; Alzheimer's disease; biomarkers

Citation: Handen, B.; Clare, I.; Laymon, C.; Petersen, M.; Zaman, S.; O'Bryant, S.; Minhas, D.; Tudorascu, D.; Brown, S.; Christian, B., on behalf of the Alzheimer's Biomarker Consortium-Down Syndrome (ABC-DS). Acute Regression in Down Syndrome. *Brain Sci.* **2021**, 11, 1109. https://doi.org/10.3390/brainsci11081109

Academic Editor: Margaret B. Pulsifer

Received: 17 July 2021
Accepted: 13 August 2021
Published: 23 August 2021

Publisher's Note: MDPI stays neutral with regard to jurisdictional claims in published maps and institutional affiliations.

1. Introduction

There has been growing interest over the past decade in acute regression among adolescents and young adults with Down syndrome (DS) [1,2]. Characterized by a sudden, and often unexplained, reduction in expressive language, decreased functional living skills and reduced psychomotor activity; regression can result in a significant change in the long-term needs and independence of these individuals. Recent studies have identified a number of potential triggers and associations for regression, including medical issues (e.g., surgery, Hashimoto's disease, sleep apnea, sleep disruption, menarche and hormonal cycles, depression) and psychosocial stressors (e.g., transition from school, change in living arrangement) [3,4]. However, it is not understood why certain individuals with DS are at risk for regression in response to such events while the vast majority are able to cope effectively. Acute regression is reported to occur in up to 16% of individuals with DS [5] and has also been referred to by a number of other terms, including "new onset autism regression", "regression, dementia and insomnia", "catatonia" and "down syndrome degenerative disorder" [6,7]. The disorder appears to occur during adolescence through the mid 20s and can either be sudden onset or progressive.

Case reports and longitudinal studies of individuals with DS who have experienced acute regression indicate that only about 10% completely regain prior levels of functioning. Approximately 40% improve to some degree, but often fail to regain prior communication skill levels. The remainder fails to regain the skills that were lost [4]. Mircher and colleagues [4] present some of the most recent data on this disorder, describing a cohort of 30 adolescents and young adults with DS who experienced acute regression. In terms of clinical presentation, the most frequently reported psychiatric symptoms included mood disorders (30%), apathy, extreme slowness or catatonia (37%) and stereotypies (27%). Forty percent of the cohort displayed self- or other-directed aggressive behavior. Speech impairment occurred among almost the entire cohort (94%). Structural MRI scans were available for 15 patients and were found to be normal in 11 individuals. The only abnormalities noted were brain atrophy ($n = 2$) and hippocampal abnormalities ($n = 2$). In 2012, Akaloshi et al. [8] described a cohort of 13 adolescents/young adults (mean age 21.2 years) that were diagnosed with acute regression in Japan. All underwent MRI or CT scanning at the time of diagnosis and were treated and followed by the authors. Five of the 13 cases exhibited MRI or CT results that were suggestive of dementia, including mild cerebral white matter ischemia, hippocampal atrophy and basal ganglia calcification. However, similar to the data presented by Mircher et al. [4], no control scans (either of non-affected individuals with DS or neurotypical individuals) were included for comparison. In addition, many of the differences on the MRI and CT scans occur among individuals with DS in the absence of regression, and hippocampal abnormalities, in particular, have been previously documented in the DS population in comparison to neurotypical individuals [9,10]. Following a range of pharmacologic interventions, 23% of the Akaloshi et al. [8] cohort were subsequently rated as "improved", 54% as "partially improved" and 23% as "no difference".

Based upon the presentation and loss of skills, some investigators have proposed that regression in DS might be related to dementia [7]. In fact, it is not unusual for individuals with DS who experience acute regression to be given a diagnosis of Alzheimer's disease (AD) or dementia, based upon the symptom presentation and loss of skills. Consequently, practitioners may prescribe drugs, such as cholinesterase inhibitors (e.g., donepezil), which are commonly used to prevent memory loss in neurotypical adults with dementia. Actually, adults with DS are at significant risk for the development of AD, with most individuals with DS displaying the neuropathology associated with AD by 40 years of age [11,12]. This is thought to be due in large part to the presence of a third APP gene on the 21st chromosome, resulting in the accelerated production of β-amyloid (referred to as "amyloid") throughout the lifetime. Amyloid and amyloid plaques are believed to be key to initiating a cascade of subsequent events, including the hyperphosphorylation and accumulation of neurofibrillary tangles comprised of tau protein, as well as changes in brain structure and functioning (e.g., decreased grey matter density, decreased hippocampal volume, increased white matter hyperintensities) that lead to dementia.

Blood based biomarkers of amyloid peptides (amyloid beta [Aβ 40, 42]) have also been increasingly explored in adults with DS due in part to the early accumulation of this protein. Findings among those with DS and AD have been mixed, with some studies reporting elevations in Aβ 1–42 [13–16] and Aβ 1–40 [13,14,17,18] while others reporting a relative decrease [15,17,18] in levels, which corresponds with CSF findings [19,20]. Lower levels have also been noted among prodromal AD groups in comparison to healthy controls [13]. In contrast, other plasma biomarkers of AD pathology, including tauopathy (total tau) and neurodegeneration (neurofilament light chain [NfL]), have shown more consistent findings, with elevations seen among individuals with DS who have been diagnosed with AD (DS-AD) [13,15,21–23].

Finally, there is evidence of even earlier pathological changes in DS, such as increased levels of non-fibrillated amyloid in the teenage years along with non-developmental grey matter and ventricular changes [24]. However, while the symptoms of acute regression appear to mirror some of those of dementia, they differ in that they tend to be sudden rather than gradual and also are not followed by a subsequent and continued loss of skills over a 2–5 year period. Yet, it is also possible that some of the early pathological changes that are documented in adolescents and adults with DS who experience regression, might continue to play a significant role in determining an increased risk for dementia in adulthood.

The Alzheimer's Biomarker Consortium–Down Syndrome (ABC–DS) is a longitudinal study of risk factors for AD in a large cohort of adults with DS. Funded by the NIA and NICHD, the consortium has enrolled approximately 400 individuals with DS, many of whom have been followed for a number of years. A wide range of potential biomarkers of AD is collected as part of the protocol, including blood and CSF-based measures, and cognitive and adaptive functioning measures, as well as MRI and PET scans. Among the current ABC-DS cohort, five individuals who had a prior documented history of acute regression were identified. By matching them with a group of adults with DS who had not had this experience, we have a unique opportunity to examine a larger number of potential biomarker differences. Drawing on potential causes of acute regression and its possible relation to early dementia, it was hypothesized that those individuals with a history of acute regression would have an increased prevalence of risk biomarkers for AD, including greater levels of amyloid deposition, tau and brain neuropathology, and blood-based biomarkers (e.g., neurofilament light chain [NfL]) than a matched group of unaffected adults.

2. Materials and Methods

2.1. ABC-DS

The ABC-DS comprises eight university-based clinical performance sites, including the University of Pittsburgh, University of Cambridge, Harvard University, Columbia University/IBRDD, Barrow Neurological Institute, University of Wisconsin Madison, University of California Irvine and Washington University. Other sites provide additional support to the project, including the University of North Texas Health Science Center, the University of Southern California, Georgetown University, the University of Michigan and the Mayo Clinic. Study participants undergo a baseline and subsequent follow-up visits at 16-month intervals, for a total of three visits over a 32-month period.

2.2. Participants

Informed consent/assent was obtained from all participants involved in the study. The study was conducted in accordance with the Declaration of Helsinki, and the protocol was approved by the ethics committees from each of the participating universities. Medical histories of enrollees were reviewed at each clinical performance site to identify those with a history of acute regression. A total of five individuals were identified and medical histories were reviewed from the period of time during which regression purportedly occurred. A diagnosis was confirmed if there was evidence of a significant loss of communication skills and adaptive functioning. As described in Table 1, diagnoses at the time included depression ($N = 1$) and dementia ($N = 3$). Age at the time of reported regression ranged from the early to late 20s. None of the individuals returned to prior levels of functioning. Current estimated mental ages ranged from <2 years 0 months to 6 years 5 months, based upon the PPVT4 [25], a measure of receptive vocabulary. Those initially diagnosed with dementia have since had their diagnoses removed.

The five individuals with histories of regression were matched with 15 other ABC-DS participants (without reported histories of acute regression) based on biological sex, age, site and ApoE status. None of the individuals with regression, or the matched participants, had current diagnoses of AD based upon a consensus conference (comprised of study staff, a physician and a psychologist who had participated in the study visit) that included a review of neuropsychological assessment battery results, caregiver-completed questionnaires on adaptive functioning, behavioral concerns and possible symptoms of AD. In addition, the consensus conference members had access to each participant's medical history and the results of a physical/neurological examination conducted as part of their study visit.

Consensus conference members were blinded as to neuroimaging, omics and genetics results. Some individuals were given diagnoses of "unable to determine" due to the likely possibility that other factors (e.g., a recent illness, change in living situation or job) might have accounted for any reported changes in overall functioning. ApoE carrier status was obtained and karyotyping was conducted to confirm the trisomy 21 diagnosis. Table 2 provides demographic information for the "regression group" and the 15 "matched controls". The only meaningful difference between the two groups was estimated mental age, with the regression group having a mean MA considerably lower than the comparison group.

2.3. Dependent Measures

Methods describing the neuroimaging and omics analyses have been published previously [26]. Structural MRI scans were obtained for all participants using a 3T MRI system with T1-weighted pulse sequences. The following scanners were used: GE Discovery MR750, Siemens Prisma and GE Signa PET/MR. The structural MRI was processed with Freesurfer (version 5.3) to determine hippocampal, caudate and putamen volumes. Both the $A\beta$ and tau PET scans were conducted with all participants, using nominal injections of 15 mCi of [^{11}C]PiB and 10 mCi of [^{18}F]AV-1451 which were administered as 20–30 s bolus injections followed by a saline flush (one site used lower doses of both imaging agents). Tracer concentration images were generated from 50–70 min post injection for [^{11}C]PiB and 80–100 min post injection for [^{18}F]AV-1451. Each subject's PET images were registered to the corresponding T1 image using PMOD.

To provide an amyloid index that could be compared with other populations (e.g., late-onset Alzheimer's disease, early-onset autosomal dominant Alzheimer's disease), the [^{11}C]PiB uptake was quantified on a universal centiloid scale using the procedure described by Klunk et al. [27]. Briefly, subject T1 MR images along with the [^{11}C]PiB images were warped to the Montreal Neurological Institute (MNI)-152 T1-weighted template using SPM8. The warped [^{11}C]PiB images were sampled using the centiloid cortex volume of interest (CTX VOI) and normalized using activity sampled with the centiloid whole cerebellum reference region volume of interest (WC VOI). Both VOIs are available at Global Alzheimer's Association Information Network (GAAIN; http://www.gaain.org (accessed on 22 January 2018)). The determined SUVR was then converted to a centiloid value using equation Equation 1.3b of [27]. Individuals with a centiloid value greater than 22 were considered to be "amyloid positive" [28].

For evaluation of AV-1451 uptake, a probability template method [29] was employed. In this procedure, atlases are defined for a number of template images which are then warped on to the subject's MR. In the current study, MR images from twelve individuals were used as templates/atlases. The template MRI images were selected through a review of images that had been processed through FreeSurfer 5.3, software which automatically parcellates brain MR images and produces a native space version of the Desikan–Killiany (DK) atlas [30]. Selection of the 12 images for use as templates was based on the quality of the parcellation results. For each participant in this study all 12 template images and corresponding atlases were warped to the participant's T1 MR, resulting in 12 versions of each DK region. Each final DK region for the participant was taken to be the volume of maximum overlap of the 12. The final DK regions were used to construct the six Braak [31] regions described in Schöll et al. [32] (with the exception that the striatum was not included in the Braak region 5). The uptake of AV-1451 in each of the Braak regions was quantified as regional SUVR, i.e., the activity concentration in the region normalized to cerebellar gray matter activity concentration.

Table 1. Description of participants.

Part	Sex	Current Age	Current IQ/MA	ApoE Status	Age at Regression	Description of Regression
01	M	33	6 years 5 months	E3/E3	28	Individual had some loss of functioning, primarily deterioration in verbal skills and increased irritability. They were given a diagnosis of early dementia at the time. Some improvement occurred over the following 2 years, although not to previous levels of functioning. Dementia diagnosis was subsequently removed.
02	F	44	4 years 0 months	E3/E3	27–28	Individual used to have a part-time job and spoke fluently. After regression, was unable to perform her job and no longer traveled independently. b They also lost interest in most activities and stopped keeping up with friends. No diagnosis of dementia was made. There has been gradual improvement since that time, although not to previous levels of functioning.
03	F	36	3 years 8 months	E3/E3	Mid 20s	Individual was given a dementia diagnosis approximately 10 years ago following a rapid decline in functioning. They were prescribed donepezil and improved a little. The dementia diagnosis was subsequently removed. They continued to improve after medication was discontinued but did not return to prior levels of functioning.
04	M	54	3 years 1 mon.	E4/E3	Early 20s.	Individual had a significant loss of functioning in their early 20s. Prior to regression they were highly verbal and independent. Subsequently, they were unable to talk, complete basic self-care activities, or participate in work and social activities. They were diagnosed with depression and prescribed an SSRI. There was some minimal improvement in language subsequently noted along with some improvement in self-care and social skills. They have maintained this same level of functioning since then and continue to have a diagnosis of depression and to take medication. They were never officially diagnosed with regression.
05	F	28	<2 years 0 months	E3/E3	Mid 20s	Individual evidenced significant loss of skills in early 20s and was diagnosed with dementia. Some small gains have been noted since that time, but not to prior levels of functioning. The dementia diagnosis was removed, however, significant behavioral issues continue along with limited language.

Table 2. Characteristics of regression and matched groups.

Variable	Regression Group	Matched Group 1:15
Age [Mean (SD)/N(%)]	38.06 (8.78)	38.01 (8.06)
Mental Age	3.42 (2.30)	8.16 (4.40)
Sex (Male)	2 (40%)	6 (40%)
ApoE Allele *	1 (20%)	3 (20%)
Karyotype		
Full Trisomy 21	4 (80%)	12 (80%)
Partial Trisomy	1 (20%)	3 (20%)
Cognitive Level		
Mild	0 (0%)	4 (26.7%)
Moderate	3 (60%)	11 (73.3%)
Severe	2(40%)	0 (0%)
Consensus Diagnosis		
Cognitively Stable	4 (80.0%)	13 (86.7%)
Unable to Determine	1 (20.0%)	2 (13.3%)

* ApoE4 = apolipoprotein E gene variant 4.

Blood samples were obtained concurrently with the PET scans. Plasma samples were assessed at the University of North Texas Health Science Center (UNTHSC) Institute for Translational Research (ITR) Biomarker Core using a single molecule array technology (Simoa; Quanterix, Billerica, MA, USA). Commercially available kits from Quanterix were utilized. Samples were loaded onto a 96-well plate and analyzed on the Simoa HD-1. Plasma concentrations of Aβ 1–42 and Aβ 1–40, total tau and NfL were obtained for most participants. The UNTHSC ITR Biomarker Core has assayed >5000 on the Simoa platform with coefficients of variability (CVs) <4%. Lower Aβ 1–42 and Aβ 1–40 values and higher total tau and NfL values would be predicted when comparing individuals with regression and unaffected controls.

Statistical Analysis: Due to the small number of individuals identified with histories of regression, only descriptive statistics are provided: means and standard deviations for the continuous measures and counts/frequencies for the dichotomous or non-interval type. A 1:1 and a 1:3 matching were performed based on several variables (age, ApoE status, karyotype and gender).

3. Results

The five participants with histories of regression were matched with 15 unaffected adults with DS. There did not appear to be any meaningful differences between the groups on demographic variables of interest, with the exception of estimated mental age. The MRI for one member of the regression group was unable to be interpreted, due to extensive movement. As a result, data were not available for any of the MRI variables for this individual. The results of the probability template method produced unsatisfactory results for one participant, with the consequence that no tau results were available. As shown in Table 3, there does not appear to be a clinically meaningful difference between groups on mean hippocampal thickness (the two groups differ by less than 2%). Conversely, the regression group actually has slightly larger right and left mean hippocampal, caudate and putamen volumes than the unaffected DS controls. Mean global centiloid SUVR (a measure of brain amyloid) appears to be slightly lower in the regression group than the unaffected controls. However, mean tau PET SUVr units across all six Braak regions suggest minimal differences between the regression group and the 15-member unaffected group.

Potentially meaningful differences between groups may have been found on some of the proteomics measures (see Table 3). The mean values of both Aβ40 and Aβ42 were slightly lower for those with histories of regression versus unaffected controls (8.7% and 4.3% lower, respectively). As a result, the mean Aβ40/Aβ42 ratio was similarly impacted (with the regression group mean ratio being 4.9% lower than the unaffected control group mean ratio). Potentially meaningful differences were also noted on for both total tau and NfL. The regression group mean NfL value was 37.0% higher than the unaffected controls' mean NfL values. Similarly, the regression group mean total tau value was 39.3% higher than the unaffected controls' mean total tau values.

Table 3. Biomarkers.

	Regress. Group (N)	Mean (SD)	Median	Matched Group 1:15 (N)	Mean (SD)	Median
MRI Scan						
Hippocampal Volume						
Left	4	3625.3 (283.8)	3618.4	11	3174.7 (435.2)	3163.2
Right	4	3655.1 (299.2)	3653.3	11	3156.3 (498.0)	3120.9
Hippocampal Thickness	4	3.00 (0.11)	2.98	11	2.95 (0.21)	2.86
Caudate						
Left	4	3471.6 (415.4)	3372.6	11	3219.0 (579.6)	3256.4
Right	4	3505.3 (475.5)	3447.4	11	3310.7 (375.8)	3352.0
Putamen						
Left	4	6205.7 (453.7)	6375.3	11	5408.4 (748.4)	5149.4
Right	4	6031.2 (356.6)	5903.3	11	5523.5 (578.2)	5300.0
Amyloid PET						
Centiloid SUVR	5	17.77 (22.09)	6.37	15	20.70 (22.59)	18.66
Amyloid Negative $\leq$ 22		3 (60%)			9 (60%)	
Amyloid Positive >= 22		2 (40%)			6 (40%)	
TAU PET						
Braak 1	4	1.22 (0.16)	1.20	15	1.19 (0.19)	1.14
Braak 2	4	1.17 (0.26)	1.12	15	1.18 (0.17)	1.11
Braak 3	4	1.11 (0.09)	1.09	15	1.13 (0.14)	1.09
Braak 4	4	1.08 (0.08)	1.10	15	1.11 (0.12)	1.08
Braak 5	4	1.08 (0.04)	1.10	15	1.09 (0.13)	1.07
Braak 6	4	1.06 (0.04)	1.08	15	1.03 (0.07)	1.02
Plasma						
Aβ40 pg/mL	4	413.25 (52.51)	403	14	452.43 (86.48)	452
Aβ42 pg/mL	4	15.13 (2.89)	14.60	14	15.81 (3.37)	15.60
Aβ40/ Aβ42 ratio	4	27.64 (2.93)	27.43	14	29.06 (4.34)	29.25
NfL pg/mL	5	18.23 (11.37)	15.20	15	11.49 (6.51)	10.80
Total Tau pg/mL	4	6.13 (6.55)	3.26	14	3.72 (4.09)	2.68

4. Discussion

This study sought to examine the possibility that early regression in adults with DS might lead to an increased risk for subsequent AD in later life, resulting in AD symptoms occurring at an earlier age than among unaffected adults with DS. Using data from the ABC-DS study, we were able to identify five individuals with histories of regression during their early to late 20s and to match them with 15 unaffected individuals with DS. While none of the individuals are yet displaying signs of AD, the ABC-DS database allows us to examine the possibility that this group of individuals, with a prior history of regression, might be at increased AD risk based upon a range of AD biomarkers. Hence it was hypothesized that in comparison to matched, unaffected individuals, those with histories of regression would have increased biomarker risk measures for AD, including greater levels of amyloid deposition, tau and brain neuropathology, and blood-based biomarkers (e.g., NfL). While a small N only allowed for the presentation of descriptive statistics, results suggest some clinically meaningful differences between the two groups that could provide preliminary evidence to support this hypothesis.

MRI findings were, in fact, the opposite of what had been hypothesized, with mean hippocampal caudate and putamen volumes being slightly higher in the regression group. Prior research in the adult DS population indicates that hippocampal volume is smaller than in cognitively normal adults and that changes in hippocampal volume among adults with DS are not likely to be seen until after the age of 50 (which also is when many individuals develop dementia) [33]. Similarly, our own prior research found no significant differences in right and left hippocampal volumes when comparing individuals with DS who were amyloid positive versus those who were amyloid negative (neither group had dementia) [34]. Hence, the differences found between the two groups in the current study are likely inconsequential. However, it should also be noted that cortical thickness may be a proxy measure of inflammation and hippocampal volume may be mirroring that. In DS and in autosomal dominant AD, the cortex is at first thicker in areas which are typically affected by AD and these regions then atrophy as disease progresses [35–37].

While no differences on tau PET findings were noted, this may have been expected given the mean age of the two groups (38 years). One might not have anticipated high tau PET SUVr values until the middle to late 40s in the DS population. However, there was a difference found on our measure of brain amyloid, with the unaffected group having a 14% greater mean global centiloid SUVR than the regression group (suggesting greater amyloid burden in the former group). This finding was counter to what had been hypothesized. To provide some context for determining if such a difference might have possible clinical significance, we examined findings from our prior research comparing amyloid positive versus amyloid negative adults with DS. In a 2017 paper examining change in amyloid load in a cohort of 52 non-demented adults with DS, those who were amyloid positive (determined via PET scan) had a mean SUVR that was 32.8% higher than those who were amyloid negative [38]. Hence, a 14% difference on mean centiloid SUVR likely has little or no clinical significance in this case.

Some potential group differences were noted on the proteomics measures. For example, both plasma Aβ40 and Aβ42 were slightly lower in the regression group, which is consistent with prior plasma (15, 17–18) and CSF findings (19–20) that show declines among those with DS-AD, reflecting a similar change among those who regressed. In addition, similar to prior work (13, 15, 21–23), we found considerably higher levels of total tau and NfL, reflecting increased neurodegeneration and further AD specific pathological changes among this group. We compared these findings with those of other ABC-DS papers which examined proteomics differences between individuals who were clinically stable (CS) and those determined to have MCI or AD (based upon a consensus conference decision). In a paper on proteomic profiles in adults with DS, Petersen et al. [39] noted a 6.8% difference between the CS and MCI groups and a 7.6% difference between the CS and AD groups on mean Aβ40. A 3.9% difference between the CS and MCI groups and a 9.1% difference between the CS and AD groups were reported for mean Aβ42 (with Aβ40 and Aβ42 values being lower for the MCI and AD groups). Hence, the mean differences observed between our regression group and our unaffected control group is in this same general range. In a second paper by Petersen et al. [22], a difference of approximately 12% between the CS and MCI groups on mean NfL and a 25% difference between the CS and AD groups on mean total tau was found. It should be noted that in both papers, the MCI and AD groups were significantly older than the CS group (and that many AD biomarkers in adults with DS appear to be significantly impacted by age). As the current regression group and unaffected control group did not differ significantly in age this cannot account for these findings. Hence, the findings suggest greater proteomics biomarker risk in the regression group, reflecting increased AD pathology of amyloid deposition, tau and neurodegeneration. It is possible that, similar to changes in CSF measures of Aβ40, Aβ42, NfL and total tau occur prior to actual detectable changes on amyloid and/or tau PET scans.

As far as we are aware, this study is the first to make use of amyloid/tau PET scans and proteomics in the examination of individuals with DS who have experienced acute regression. In contrast to prior studies that have reported some MRI results, we were able to include a comparison set of scans for a matched group of unaffected individuals. However, there are a number of weaknesses in this report. First, the N is too small to conduct an adequate statistical analysis of the data. Second, some of the affected individuals did not appear to have been given an official diagnosis of regression at the time of the event (although the description of their behavior is consistent with DS regression). Third, it is possible that this group of five individuals is not representative of the larger group of adolescents and young adults with DS who experience regression. The individuals in this report may have had a more successful recovery, which may have also impacted their risk for AD. Conversely, adults with DS and regression who experienced minimal or no recovery may represent a subgroup that is at the greatest risk for AD. However, such individuals would not have been enrolled in ABC-DS, as a minimum mental age of 36 months was required. It should also be noted that the five individuals with regression were functioning at a much lower level than the comparison group. In fact, it was not possible to create a matched group that had a similar mental age to those who had experienced regression. As a result, we chose not to compare the groups on other cognitive measures, as the possibility could not be excluded that any differences might reflect their level of functioning rather than regression. However, there is no evidence that cognitive level increases the risk for AD among individuals with DS. Therefore, it was not felt that the differences in mental age would be expected to impact the values of our biomarkers of interest.

Finally, while none of these five individuals has yet developed AD (and, in general, are below the age when one would expect to see early AD signs), it is still possible that those who have experienced regression could be more susceptible than nonaffected individuals to experiencing clinically significant changes in AD biomarkers (especially neuroimaging biomarkers) and to exhibiting earlier AD symptoms. Future research should include a larger cohort of individuals who have experienced regression (and include those functioning below a mental age of three years). In addition, other potential proteomics biomarkers, such as ptau181 and ptau217, should be included. Finally, it may require a number of years of follow-up before the potential relationship between regression and increased AD risk is clearer.

Author Contributions: Conceptualization, B.H. and I.C.; methodology, B.H. and D.T.; formal analysis, M.P., C.L., S.O., B.C. and D.M.; investigation, all authors; resources, C.L., M.P. and S.O.; data curation, M.P., D.T. and C.L.; writing—original draft preparation, B.H.; writing—review and editing, all authors; supervision, B.H., D.T. and S.O.; funding acquisition, B.H., B.C., S.Z. and S.O. All authors have read and agreed to the published version of the manuscript. The authors present this work on behalf of the Alzheimer's Biomarker Consortium-Down Syndrome (ABC-DS).

Funding: This research was funded by the National Institute of Aging and the National Institute on Child Health and Human Development (U01AG051406; U19AG068054).

Institutional Review Board Statement: The study was conducted according to the guidelines of the Declaration of Helsinki and approved by the Institutional Review Board of the University of Pittsburgh (PRO15100180; 25 January 2016), Barrow Neurological Institute (PHX-16-0044-11-03; 4 December 2016), University of Wisconsin, Madison (2016-0738; 3 October 2016) and University of Cambridge (16/YH/0297; 3 February 2017).

Informed Consent Statement: Informed consent/assent was obtained from all subjects involved in the study.

Data Availability Statement: Requests for qualified investigators to obtain data supporting the reported results can be made at https://pitt.co1.qualtrics.com/jfe/form/SV_cu0pNCZZlrdSxUN, accessed on 13 August 2021.

Acknowledgments: The authors would like to thank the project coordinators (Cathleen Wolfe, Sandy Quintanilla, Monika Grigorova and Jessica Beresford-Webb) as well as our global coordinator (Joni Vander Bilt) who made this research possible. We would also like to thank Laisze Lee for all her contributions in terms of statistical summaries and data management. Finally, we would like to thank the adults with Down syndrome and their families for their time and commitment to further discovery and understanding into the causes of Alzheimer's disease.

Conflicts of Interest: The authors declare no conflict of interest. The funders had no role in the design of the study; in the collection, analyses, or interpretation of data; in the writing of the manuscript or in the decision to publish the results.

References

1. Rosso, M.; Fremion, E.; Santoro, S.L.; Dreskovic, N.M.; Chitnis, T.; Skotko, B.G.; Santoro, J.D. Down syndrome disintegrative disorder: A clinical regression syndrome of increasing importance. *Pediatrics* **2019**, *1456*, 2019–2939. [CrossRef]
2. Walpert, M.; Holland, A.; Zaman, S. A systematic review of unexplained early regression in adolescents and adults with Down's syndrome. *Brain Sci.* **2021**. in submission.

3. Devenny, D.; Matthews, A. Regression: Atypical loss of attained functioning in children and adolescents with Down syndrome. *Int. J. Dev. Disabil.* **2011**, *41*, 233–264.
4. Mircher, C.; Cieuta-Walti, C.; Marey, I.; Rebillat, A.S.; Cretu, L.; Milenko, E.; Conte, M.; Sturtz, F.; Rethore, M.O.; Ravel, A. Acute regression in young people with Down syndrome. *Brain Sci.* **2017**, *7*, 57. [CrossRef]
5. Ghazuiddin, N.; Nassiri, A.; Miles, J.H. Catatonia in Down syndrome; a treatable cause of regresson. *Neuropsychiatr. Dis. Treat.* **2015**, *11*, 941–949. [CrossRef] [PubMed]
6. Santoro, S.L.; Cannon, S.; Capone, G. Unexplained regression in Down syndrome: 35 cases from an international Down syndrome database. *Genet. Med.* **2020**, *22*, 767–776. [CrossRef]
7. Worley, G.; Crissman, B.G.; Cadogan, E.; Milleson, C.; Adkins, D.W. Down syndrome disintegrative disorder: New-onset autistic regression, dementia, and insomnia in older children and adolescents with Down syndrome. *J. Child Neurol.* **2015**, *30*, 1147–1152. [CrossRef]
8. Akahoshi, K.; Matsuda, H.; Funahashi, M.; Hanaoka, T.; Suzuki, Y. Acute neuropsychiatric disorders in adolescents and young adults with Down syndrome: Japanese case reports. *Neuropsychiatr. Dis. Treat.* **2012**, *8*, 339–345. [CrossRef] [PubMed]
9. Fortea, J.; Vilaplana, E.; Carmona-Iragui, M.; Benejam, B.; Videla, L.; Barroeta, I.; Fernández, S.; Altuna, M.; Pegueroles, J.; Montal, V.; et al. Clinical and biomarker changes of Alzheimer's disease in adults with Down syndrome: A cross-sectional study. *Lancet* **2020**, *395*, 1988–1997. [CrossRef]
10. Teipel, S.J.; Schapiro, M.B.; Alexander, G.E.; Krasuski, J.S.; Horwitz, B.; Hoehne, C.; Moller, H.J.; Rapoport, S.; Hampel, H. Relation of corpus callosum and hippocampal size to age in nondemented adults with Down's syndrome. *Am. J. Psychiatry* **2003**, *160*, 1870–1878. [CrossRef] [PubMed]
11. Wiseman, F.K.; Al-Janabi, T.; Hardy, J.; Karmiloff-Smith, A.; Nizetic, D.; Tybulewicz, V.L.; Fisher, E.M.; Strydom, A. A genetic cause of Alzheimer disease: Mechanistic insights from Down syndrome. *Nat. Rev. Neurosci.* **2015**, *16*, 564–574. [CrossRef] [PubMed]
12. Zigman, W.B.; Devenny, D.A.; Krinsky-McHale, S.J.; Jenkins, E.C.; Urv, T.K.; Wegiel, J.; Schupf, N.; Silverman, W. Alzheimer's disease in adults with Down syndrome. *Int. Rev. Res. Ment. Retard.* **2008**, *36*, 103–145. [PubMed]
13. Fortea, J.; Carmona-Iragui, M.; Benejam, B.; Videla, L.; Barroeta, I.; Fernández, S.; Altuna, M.; Pegueroles, J.; Montal, V.; Valldeneu, S.; et al. Plasma and CSF biomarkers for the diagnosis of Alzheimer's disease in adults with Down syndrome: A cross-sectional study. *Lancet Neurol.* **2018**, *17*, 860–869. [CrossRef]
14. Iulita, M.F.; Ower, A.; Barone, C.; Pentz, R.; Gubert, P.; Romano, C.; Cantarella, R.A.; Elia, F.; Buono, S.; Recupero, M.; et al. An inflammatory and trophic disconnect biomarker profile revealed in Down syndrome plasma: Relation to cognitive decline and longitudinal evaluation. *Alzheimer's Dement.* **2016**, *12*, 1132–1148. [CrossRef]
15. Lee, N.C.; Yang, S.Y.; Chieh, J.J.; Huang, P.T.; Chang, L.M.; Chiu, Y.N.; Huang, A.C.; Chien, Y.H.; Hwu, W.L.; Chiu, M.J. Blood beta-amyloid and tau in down syndrome: A comparison with Alzheimer's disease. *Front. Aging Neurosci.* **2017**, *8*, 316. [CrossRef]
16. Schupf, N.; Patel, B.; Pang, D.; Zigman, W.B.; Silverman, W.; Mehta, P.D.; Mayeux, R. Elevated plasma β-amyloid peptide Aβ42 levels, incident dementia, and mortality in Down syndrome. *Arch. Neurol.* **2007**, *64*, 1007–1013. [CrossRef]
17. Coppus, A.M.; Schuur, M.; Vergeer, J.; Janssens, A.C.; Oostra, B.A.; Verbeek, M.M.; van Duijn, C.M. Plasma beta amyloid and the risk of Alzheimer's disease in Down syndrome. *Neurobiol. Aging* **2012**, *33*, 1988–1994. [CrossRef] [PubMed]
18. Head, E.; Doran, E.; Nistor, M.; Hill, M.; Schmitt, F.A.; Haier, R.J.; Lott, I.T. Plasma amyloid-beta as a function of age, level of intellectual disability, and presence of dementia in Down syndrome. *J. Alzheimers Dis.* **2011**, *23*, 399–409. [CrossRef]
19. Tamaoka, A.; Sekijima, Y.; Matsuno, S.; Tokuda, T.; Shoji, S.; Ikeda, S.I. Amyloid beta protein species in cerebrospinal fluid and in brain from patients with Down's syndrome. *Ann. Neurol.* **1999**, *46*, 933. [CrossRef]
20. Tapiola, T.; Soininen, H.; Pirttila, T. CSF tau and Abeta42 levels in patients with Down's syndrome. *Neurology* **2001**, *56*, 979–980. [CrossRef] [PubMed]
21. Kasai, T.; Tatebe, H.; Kondo, M.; Ishii, R.; Ohmichi, T.; Yeung, W.T.E.; Morimoto, M.; Chiyonobu, T.; Terada, N.; Allsop, D.; et al. Increased levels of plasma total tau in adult Down syndrome. *PLoS ONE* **2017**, *12*, e0188802. [CrossRef]
22. Petersen, M.E.; Rafii, M.S.; Julovich, D.; Zhang, F.; Hall, J.; Ances, B.M.; Schupf, N.; Krinsky-McHale, S.J.; Mapstone, M.; Silverman, W.; et al. Plasma total-tau and neurofilament light chain as diagnostic biomarkers of Alzheimer's disease dementia and mild cognitive impairment in adults with Down syndrome. *Alzheimer's Dement. Diagn. Assess. Dis. Monit.* **2021**, *79*, 671–681.
23. Strydom, A.; Heslegrave, A.; Startin, C.M.; Mok, K.Y.; Hardy, J.; Groet, J.; Nizetic, D.; Zetterberg, H.; LonDownS Consortium. Neurofilament light as a blood biomarker for neurodegeneration in Down syndrome. *Alzheimer's Res. Ther.* **2018**, *10*, 39. [CrossRef]
24. Koran, M.E.; Hohman, T.J.; Edwards, C.M.; Vega, J.N.; Pryweller, J.R.; Slosky, L.E.; Crockett, G.; de Rey, L.V.; Meds, S.A.; Danker, N.; et al. Differences in age-related effects on brain volume in Down syndrome as compared to Williams' syndrome and typical development. *J. Neurodev. Disord.* **2014**, *6*, 1–11. [CrossRef] [PubMed]
25. Dunn, L.M.; Dunn, D.M. *Peabody Picture Vocabulary*, 4th ed.; NCD Pearson, Inc.: San Antonio, TX, USA, 2007.
26. Handen, B.L.; Lott, I.T.; Christian, B.; Schupf, N.; O'Bryant, S.; Mapstone, M.; Fagan, A.M.; Lee, J.; Tudorascu, D.; Wang, M.; et al. The Alzheimer's Biomarker Consortium-Down Syndrome (ABC-DS): Rationale and methodology. *Alzheimer's Dement. Diagn. Assess. Dis. Monit.* **2020**, *12*, 1–15.
27. Klunk, W.E.; Koeppe, R.A.; Price, J.C.; Benzinger, T.L.; Devous, M.D.; Jagust, W.J.; Johnson, K.A.; Mathis, C.A.; Minhas, D.; Pontecorvo, M.J.; et al. The Centiloid Project: Standardizing quantitative amyloid plaque estimation by PET. *Alzheimers* **2015**, *11*, 1–15. [CrossRef] [PubMed]

28. Knopman, D.S.; Lundt, E.S.; Therneau, T.M.; Albertson, S.M.; Gunter, J.L.; Senjem, M.L.; Schwarz, C.G.; Mielke, M.M.; Machulda, M.M.; Alzheimer's Dis Neuroimaging Initiative; et al. Association of Initial beta-Amyloid Levels with Subsequent Flortaucipir Positron Emission Tomography Changes in Persons Without Cognitive Impairment. *JAMA Neurol.* **2021**, *78*, 217–228. [CrossRef]
29. Svarer, C.; Madsen, K.; Hasselbalch, S.G.; Pinborg, L.H.; Haugbol, S.; Frokjaer, V.G.; Holm, S.; Paulson, O.B.; Knudsen, G.M. MR-based automatic delineation of volumes of interest in human brain PET images using probability maps. *Neuroimage* **2005**, *24*, 969–979. [CrossRef] [PubMed]
30. Desikan, R.S.; Ségonne, F.; Fischl, B.; Quinn, B.T.; Dickerson, B.C.; Blacker, D.; Buckner, R.L.; Dale, A.M.; Maguire, R.P.; Hyman, B.T.; et al. An automated labeling system for subdividing the human cerebral cortex on MRI scans into gyral based regions of interest. *Neuroimage* **2006**, *31*, 968–980. [CrossRef]
31. Braak, H.; Braak, E. Neuropathological staging of alzheimer-related changes. *Acta Neuropathol.* **1991**, *82*, 239–259. [CrossRef]
32. Scholl, M.; Lockhart, S.N.; Schonhaut, D.R.; O'Neil, J.P.; Janabi, M.; Ossenkoppele, R.; Baker, S.L.; Vogel, J.W.; Faria, J.; Schwimmer, H.D.; et al. PET Imaging of Tau Deposition in the Aging Human Brain. *Neuron* **2016**, *89*, 971–982. [CrossRef] [PubMed]
33. Aylward, E.H.; Li, Q.; Honeycutt, N.A.; Warren, A.C.; Pulsifer, M.B.; Barta, P.E.; Chan, M.D.; Smith, P.D.; Jerram, M.; Pearlson, G. MRI volumes of the hippocampus and amygdala in adults with Down's syndrome with and without dementia. *Am. J. Psychiatry* **1999**, *156*, 564–568. [PubMed]
34. Hartley, S.L.; Handen, B.L.; Devenny, D.A.; Hardison, R.; Mihaila, I.; Price, J.C.; Cohen, A.D.; Klunk, W.E.; Mailick, M.R.; Johnson, S.C.; et al. Cognitive functioning in relation to the accumulation of brain β-amyloid in healthy adults with Down syndrome. *Brain* **2014**, *137*, 2556–2563. [CrossRef] [PubMed]
35. Montal, V.; Vilaplana, E.; Peguerloes, J.; Bejanin, A.; Alcolea, D.; Carmona-Iragui, M.; Clarimón, J.; Levin, J.; Cruchaga, C.; Graff-Radford, N.R.; et al. Biphasic cortical macro- and microstructural changes in autosomal Alzheimer's disease. *Alzheimers* **2021**, *17*, 618–628. [CrossRef] [PubMed]
36. Vilaplana, E.; Rodriguez-Vieitz, E.; Ferreira, D.; Montal, V.; Almkvist, O.; Wall, A.; Lleó, A.; Westman, E.; Graff, C.; Fortea, J.; et al. Cortical microstructural correlates of astrocytosis in autosomal-dominant Alzheimer disease. *Neurology* **2020**, *94*, e2026–e2036. [CrossRef] [PubMed]
37. Annus, T.; Wilson, L.R.; Acosta-Cabronero, J.; Cardenas-Blanco, A.; Hong, Y.T.; Fryer, T.D.; Coles, J.P.; Menon, D.K.; Zaman, S.H.; Holland, A.J.; et al. The Down syndrome brain in the presence and absence of fibrillar β-amyloidosis. *Neurobiology* **2017**, *53*, 11–19. [CrossRef]
38. Lao, P.J.; Handen, B.L.; Betthauser, T.J.; Mihaila, I.; Hartley, S.L.; Cohen, A.D.; Tudorascu, D.L.; Bulova, P.D.; Lopresti, B.J.; Tumuluru, R.V.; et al. Longitudinal changes in amyloid PET and volumetric MRI in the non-demented Down syndrome population. *Alzheimer's Dement. Diagn. Assess. Dis. Monit.* **2017**, *9*, 1–9.
39. Petersen, M.E.; Zhang, F.; Schupf, N.; Krinsky-McHale, S.J.; Hall, J.; Mapstone, M.; Cheema, A.; Silverman, W.; Lott, I.T.; Rafii, M.S.; et al. Proteomic Profiles for Alzheimer's Disease and Mild Cognitive Impairment Among Adults with Down Syndrome Spanning Serum and Plasma: An ABC-DS Study. the Alzheimer's Biomarker Consortium—Down Syndrome (ABC-DS). *Alzheimer's Dement. Diagn. Assess. Dis. Monit.* **2020**, *12*, 1–11. [CrossRef]

Systematic Review

A Systematic Review of Unexplained Early Regression in Adolescents and Adults with Down Syndrome

Madeleine Walpert, Shahid Zaman and Anthony Holland *

Cambridge Intellectual and Developmental Disabilities Research Group, Department of Psychiatry, University of Cambridge, Cambridge CB2 8AH, UK; CIDDRG@medschl.cam.ac.uk (M.W.); shz10@medschl.cam.ac.uk (S.Z.)
* Correspondence: ajh1008@medschl.cam.ac.uk

Abstract: A proportion of young people with Down syndrome (DS) experience unexplained regression that severely impacts on their daily lives. While this condition has been recognised by clinicians, there is a limited understanding of causation and an inconsistent approach to diagnosis and treatment. Varied symptomology and little knowledge of the cause of this regression have impacted on clinician's ability to prevent or manage this condition. The purpose of this review was to examine the current evidence surrounding unexplained regression in adolescents and young adults, and to establish patterns that may be of use to clinicians, as well as raising awareness of this condition. Four areas were specifically reviewed, (1) terminology used to refer to this condition, (2) the symptoms reported, (3) potential trigger events and, (4) treatments and prognosis. A variety of terminology is used for this condition, which has constrained past attempts to identify patterns. An extensive number of symptoms were reported, however sleep impairment, loss of language and distinct changes in personality and behaviour, such as disinterest and withdrawal, were among the most frequently seen. Life events that were tentatively associated with the onset of a regressive period included a significant change in environmental circumstances or a transition, such as moving home or leaving school. Prognosis for this condition is relatively positive with the majority of individuals making at least a partial recovery. However, few patients were found to make a full recovery to their previous level of functioning and serious adverse effects could persist in those who have made a partial recovery. This is an under-researched condition with significant impacts on people with DS and their families. There are no established treatments for this condition and there is relatively little recognition in the research community. Further studies that focus on the prevention and treatment of this condition with controlled treatment trials are needed.

Keywords: Down syndrome; early regression; idiopathic regression

Citation: Walpert, M.; Zaman, S.; Holland, A. A Systematic Review of Unexplained Early Regression in Adolescents and Adults with Down Syndrome. *Brain Sci.* **2021**, *11*, 1197. https://doi.org/10.3390/brainsci11091197

Academic Editor: Margaret B. Pulsifer

Received: 12 July 2021
Accepted: 8 September 2021
Published: 10 September 2021

Publisher's Note: MDPI stays neutral with regard to jurisdictional claims in published maps and institutional affiliations.

1. Introduction

Down syndrome (DS) is the most common syndrome associated with the presence of an intellectual disability, affecting approximately 3.3–6.7 per 10,000 individuals worldwide [1]. Family members and clinicians have noted the occurrence of cognitive deterioration and skills loss specifically in a small portion of adolescents and young adults, often without a distinct cause. This unexplained regression is profound and has a serious impact on both the individual and their families. A great number of different terms have been used in the diagnosis of this regression. Recently, "Down syndrome disintegrative disorder" (DSDD) and idiopathic regression in DS (IRDS) have been used more frequently. The latter term, IRDS, will be used in this review.

Despite the occurrence of IRDS being well-recognised by clinicians as affecting a minority of young people with DS, research in this area is sparse [2–4]. Characteristically, the symptoms and signs of this condition include significant impacts on the person's cognitive and language functioning, their ability to perform daily tasks, a considerable loss of previously acquired daily skills, mild to severe alterations in personality and behaviour and the onset of social withdrawal. IRDS is described by families as having a profound effect on the abilities of the person with DS to live as they have previously been able to. This has a knock-on effect on members of their family and other carers, and often results in the need for major changes in their living situation and care needs.

This condition typically occurs in early adolescence to young adulthood and there are currently no confirmed causes or triggers and no consistent treatment pathways. In IRDS, presenting symptoms sometimes overlap with features of autism and dementia, however the age profiles for these other conditions are different. Autism spectrum disorder (ASD) presents in early childhood and a diagnosis of Alzheimer's disease (AD), based on the onset of clinical symptoms, is most commonly made in the fifth decade. With dementia in people with DS, the neuropathological hallmarks are often seen earlier, (around the 40s) [5,6], but a clinical diagnosis of dementia is not usually made until the patient is in their 50s. This high risk of AD for people with DS is customarily theorised to be linked to the triplication of chromosome 21 and therefore the presence of three copies of the amyloid precursor protein gene, and the resultant lifelong overproduction of the beta-amyloid (Aβ) protein [7]. Despite the high risk of AD specific to people with DS, there is minimal evidence to suggest that AD presentation occurs at the age when individuals are most likely to be affected by IRDS. Furthermore, with IRDS there is often stabilisation and/or recovery of symptoms, as opposed to when it is dementia where progression with no recovery is what is to be expected. It is generally accepted that IRDS symptoms are not a consequence of either the above conditions and should be considered as separate.

A recent paper by Santoro et al. [4] reported the findings from a retrospective chart review of 35 people with DS and regression. Using a checklist of symptoms were classified into five core "features" including: (a) adaptive functions, (b) functional and procedural memory deficits (c) motor control impairment; (d) catatonia and; (e) disturbances associated with mental ill-health. The strengths of this study included the analysis of symptomatology in a group of people with DS, who experts had agreed had unexplained regression, and the use of an agreed checklist of symptoms. Most importantly, and uniquely in this field, this study compared symptomatology and test scores of patients with an aged-matched group of people with DS with no evidence of IRDS, thus helping to validate the recorded clinical observations. The authors do not report the temporal sequence of specific clinical symptoms. However, the majority of symptoms identified in those with IRDS were not experienced in the healthy controls, the exceptions to this being the mental health categorisations and externalising behaviours (hyperactive, irritable, disruptive, agitated) where there were no significant differences between IRDS and DS groups.

During the preparation of this systematic review another review paper was published summarising reported studies of regression in people with DS [8]. This paper identified language regression, mood disturbance and new onset insomnia as being particularly common features. They proposed that there were two potential causative mechanisms, one relating to immune dysfunction, and the other being stress related. Clinically, it was argued that an extensive work-up is still required to identify possible rare causes of regression, including the co-occurrence of other genetic disorders, such as Lesch Nyhan syndrome, in which a similar regression occurs but much earlier in life. Our systematic review complements this paper, drawing in greater depth on case studies as well as reports on case series of IRDS in adolescents and young adults with DS, and examining how symptoms cluster and co-present. Our primary objective, by extending the work undertaken by Rosso et al. [8], was to help improve diagnosis by heightening awareness of this condition among clinicians and providing further details of the main characteristics and their relationships to each other. We also reflect on terminology, clinical practice and possible causation.

Our review has focussed on observations from case studies and research on IRDS in adolescents and young adults with DS. The specific aims were to identify patterns of (a) symptomology, (b) potential trigger events and, (c) prognosis, treatments and outcomes. Possible causation will be considered to highlight the need for treatment trials for this condition based on the understanding of causal mechanisms. In addition to raising awareness of the condition we highlight the importance and necessity of further research of this condition.

2. Methodology

2.1. Identification of Articles

This systematic review was conducted according to the Preferred Reporting Items for Systematic Reviews and Meta-Analyses (PRISMA) statement [9] and was accepted to the Prospero platform (registration CRD42019156614). PubMed and Scopus databases were searched in November 2019 using the following search strategy: ["regression"] and ["Down* syndrome"]. Publication date and article language were not restricted; however, a filter was applied restricting articles to those involving human participants only. PubMed search fields included title and abstract, and in Scopus, title, abstract and keywords were included. As this systematic review did not contain independent research, ethical approval was not required, and consent procedures were not applicable.

2.2. Inclusion Criteria

Articles were examined for relevance by manually screening of the titles, abstracts and the keywords included. References of the retained articles were studied for further relevant papers, which were then examined against the same criteria.

Inclusion criteria were as follows:

- Research article involving at least one individual with DS.
- Age of patient under 35 years.
- Evidence of at least one regressive period that included changes to cognition, functioning and/or behaviour and personality.
- Regression identified did not progress to a clinical diagnosis of AD.

Cases with a co-diagnosis of pre-morbid autism in DS and early regression due to autism were not included. The inclusion of an upper age restriction was necessary in order to minimise the number of persons with DS whose symptoms may have been caused by the onset of AD. Furthermore, any individual cases included in papers who fell outside of this age limit were excluded from the review. Those with pre-existing features of catatonia were not excluded from this review due to the high incidence of presentation as part of IRDS. No other restrictions, such as language or year of publication, were included.

2.3. Data Collection Process

Data sought from the articles obtained included (a) terminology used to describe the condition (e.g., regression, catatonia), (b) symptomology (e.g., change in mental state, general mental functioning, level of living skills, sleep, appetite), (c) noted trigger events (e.g., life events such as bereavement, physical illness), (d) treatments prescribed (medication, psychological interventions, etc.) and (e) outcomes. For case studies this information was extracted on an individual basis and for cohort studies at a group level in accordance with the style of the individual article. Where possible this information was collated, otherwise two groups were considered for analysis (Group A—case study participants, Group B—cohort study participants).

3. Methodology

3.1. Article Search Results

A total of 1938 articles were identified from the initial search. Due to the search term "regression" without further specification, our search was deliberately over-inclusive. It was felt that due to the inconsistencies of terminology and labelling used in referring to this condition this was a necessary step in order to capture as many related articles as possible and to achieve our aim of collating the various terminologies used. Title, keyword and abstract review eliminated the vast majority of articles, leaving 57 articles for full-text review. A further four articles were sourced from references and full-text screening with the same inclusion criteria applied to these articles was completed. Ultimately 14 articles were retained. A full-text version could not be sourced for three articles, and two others were sourced at a later date, leaving a final total of 13 articles for inclusion in this review. The search pathway is shown in Figure 1 and the final included articles are summarised in Table 1.

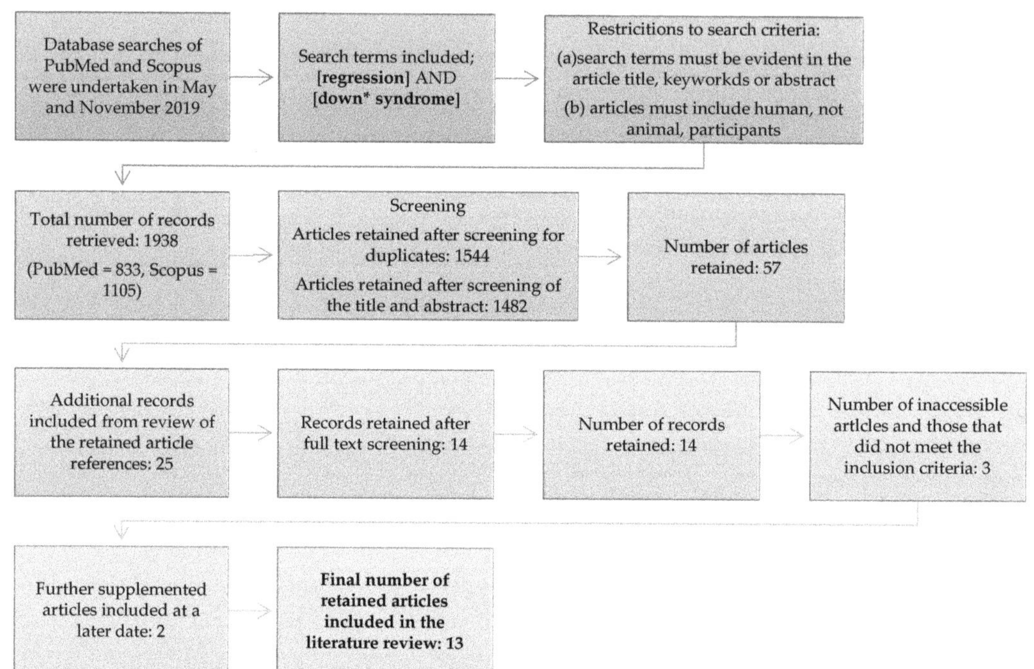

Figure 1. Conceptual framework showing article retrieval and inclusion process. Truncation was used in the search term Down* syndrome so as to capture alternate references such as Down's and Downs syndrome.

Table 1. Summary of articles included in the systematic review.

Author	Number of Case Study Patients (Group A)	Number of Cohort Study Patients (Group B)	Gender (Female: Male)	Age
Myers and Pueschel (1995) [10]	8	8	4:4	Range 21–44 years
Capone, Aidikoff and Goyal (2011) [11]	0	33	14:19	Range 13–35 years Mean 22 years
Akahoshi et al. (2012) [12]	12	12	6:6	Range 13–29 years
Stein et al. (2013) [13]	1	1	Female	13 years
Capone et al. (2013) [14]	0	28	14:14	Male mean 21.8 years Female mean 20.3 years
Dykens et al. (2015) [3]	1	49	49% male	Range 13–29 years
Ghaziuddin, Nassiri and Miles (2015) [15]	4	4	2:2	Range 14–18 years
Jacobs et al. (2016) [16]	1	1	Male	19 years
Tamasaki et al. (2016) [17]	1	1	Male	15 years
Mircher et al. (2017) [18]	0	30	20:10	Range 12–30 years
Cardinale et al. (2018) [19]	4	4	3:1	Range 17–25 years
Santoro et al. (2019) [4]	0	35	53% female	9–34 years
Miles et al. (2020) [20]	7	0	6:1	18–33 years

3.2. Additional Comments and Exclusions

Articles where the age range of the participants extended outside of our 35-year upper age limit underwent an additional level of scrutiny. In research articles where cases could not be distinguished from each other, these were excluded, however in case studies or where this information was available, only the individual cases that did not meet our inclusion criteria were excluded from that particular review. Further exclusions included specified autistic regression and progression to AD. Two individual cases were excluded from this review based on the above criteria respectively. A ten-year-old female [12] and a 44-year-old male [10]. Table 1 reflects the number of participants after these additional exclusions. Case-control studies identified that did not distinguish between young children and adolescents could not be included in the review [4,21,22].

4. Results

4.1. Patient Demographics

The total number of people with DS included in the subsequent analyses was 186. Case report data were available for 39 patients with DS, these formed Group A. Group B represented the 147 patients without individual case study reports, the cohort group. One hundred patients (53.7%) were female and 86 (46.2%) male. The mean age of onset was 20.97 years. However, this statistic is not truly representative for two reasons. First, the cohort studies provided only mean age and age range, therefore individual ages could not be entered into this analysis. Secondly, differences between age of onset and age at presentation were not always specified in the case studies. In comparison to other studies that have looked at average age of onset, our estimate may be considered high; Santoro et al., [4] for example identified 17.5 years as the typical age that regressive symptoms first appear.

4.2. Descriptive Terminology

There were considerable problems with the search strategies used in this systematic review. One of which was deliberate. The use of the word "regression" in the search terms led to massive overlap with usage in statistical terminology. This led to many retrieved articles being unrelated to the review topic. It was however necessary to include this term in order to capture all relevant articles.

One of our primary intentions of this review was to evaluate the wide variety of terminologies that are used to describe and diagnose this condition (see Table 2). Analysis of the terms used across the 13 papers entered in this review revealed that five different terms including the word "regression" were used, as well as a further 10 different descriptors not including the word regression. These terminologies have been grouped and the frequency of their appearance is noted in Table 2. Multiple terms were often used in single papers, presumably reflecting the uncertainty and inconsistency in this area.

Table 2. List of the terminology used to describe the group of patients within the article. Articles may have referenced multiple terminologies.

Regression Related Terminology	Times Used	Disorder Related Terminology	Times Used	Function Related Terminology	Times Used
Regression	2	Psychiatric disorders	1	Deterioration	1
Developmental regression	1	Down syndrome disintegrative Disorder	2	Clinical deterioration	1
Cognitive regression	1	New-onset mood disorder	1	Functional decline	1
Unexplained regression	2	Acute neuropsychiatric disorders	1		
Rapid regression	2	Depression/major depression	3		
Acute regression	2				
Total	10	Total	8	Total	3

4.3. Symptom Severity and Diversity

The majority of studies included in this review provided detailed information regarding patient symptoms. In our analysis we first assessed the qualitative data from the case studies (Group A). Our aim was to establish the most commonly reported symptoms from Group A, and then supplement these with the data from Group B. Additional symptoms that occurred in <10% of case studies reviewed were not included in this analysis or in Figure 2. Some Group B data could not be included as no symptoms were recorded [3]. In addition, symptoms were frequently reported differently between studies, for example, "catatonia" and "slowness of movement" were grouped as one category in one study [18], and "depression" and "compulsions" were grouped together in another [4]. To avoid missing the subtleties of these symptoms it was decided that, in these circumstances, patients should be recorded as having both symptoms.

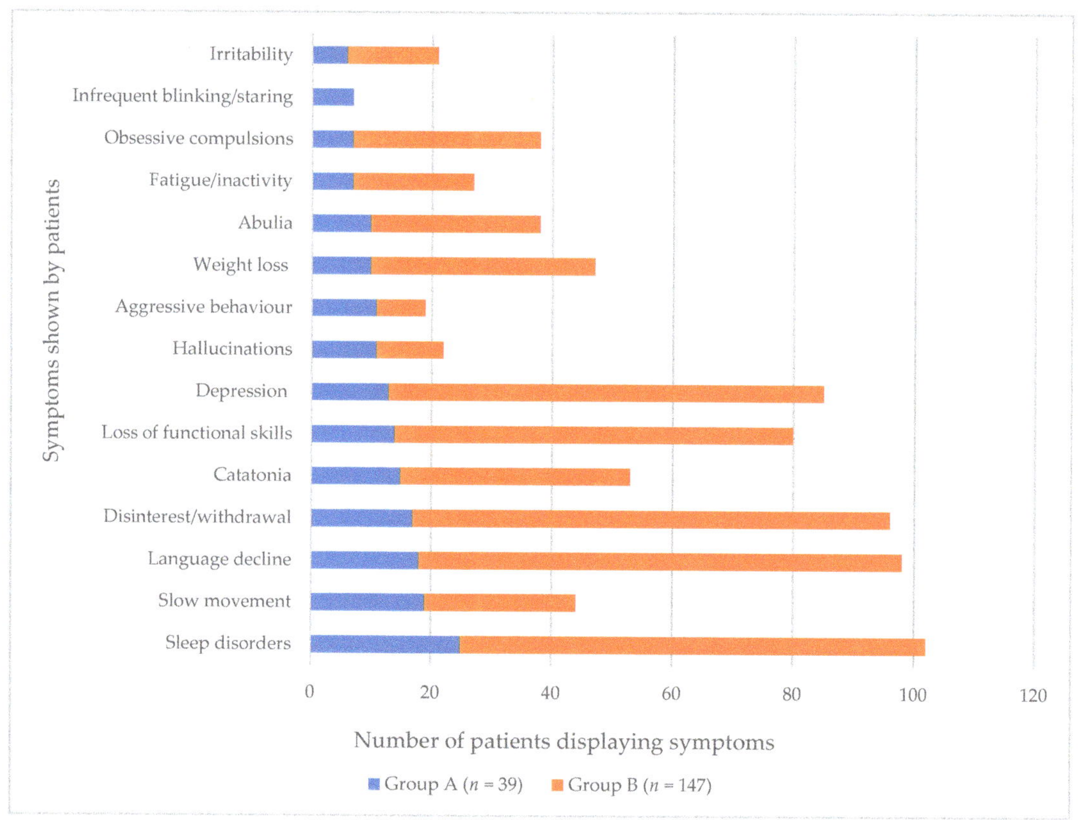

Figure 2. Number of cases in Group A (case study patients) and Group B (cohort study patients) displaying most prevalent symptoms of IRDS.

From the results of this analysis, 15 independent symptoms were identified that were present in more than 10% of Group A patients. Some symptoms were reported heavily in individual studies and yet did not feature in others. For example, abulia was included in only two papers [4,12], but featured prominently in both. In Akahoshi et al. [12] 10 of the 12 of patients included were described as showing signs of abulia and in Santoro et al. [4] 28 of 35 patients were recorded as having a symptom under the heading of "motor control", which included the features of abulia, avolition, and mutism.

Figure 2 shows the number of cases in each of the two groups of patients reported as having a particular symptom. Sleep disorders were the most commonly reported symptom in both groups and across all patients. Other highly occurring symptoms included language decline, disinterest/withdrawal, depression and loss of functional skills (self-care, toileting etc.). Onset or increase of previously present autistic characteristics were reported in three people with DS although some studies had excluded patients with co-morbid ASD [3,12]. Another symptom of particular note is weight loss, and in some cases the onset of what was described as anorexia nervosa was reported. Weight loss and poor appetite are not generally common in people with DS.

With four of the symptoms listed in Figure 2 it was possible to determine the severity based on the vocabulary used to describe the symptoms. Reviewer determined categorisation of severity of symptoms was made based on descriptions given in the case studies, shown in Table 3. Figure 3 shows the prevalence of the moderate and severe symptoms identified in the case study data (Group A patients). It was not possible to complete this analysis in the cohort studies (Group B) due to the grouping of many of symptoms together which was not consistent across studies.

Table 3. Descriptive terminology used in articles reviewed. Four of the symptoms identified from the case study data (Group A patients) were able to be analysed. Reviewer determined categorisation of "moderate" or "severe" impact.

Symptom	Moderate Symptoms	Severe Symptoms
Sleep	Restless sleep Poor sleep Disturbed sleep	Insomnia
Language	Vocal stereotypies Language decline Incoherent speech	Mutism
Weight loss	Weight loss Appetite loss	Anorexia nervosa
Slowing of movement	Slowness Slow movement	Immobility Becoming bedridden

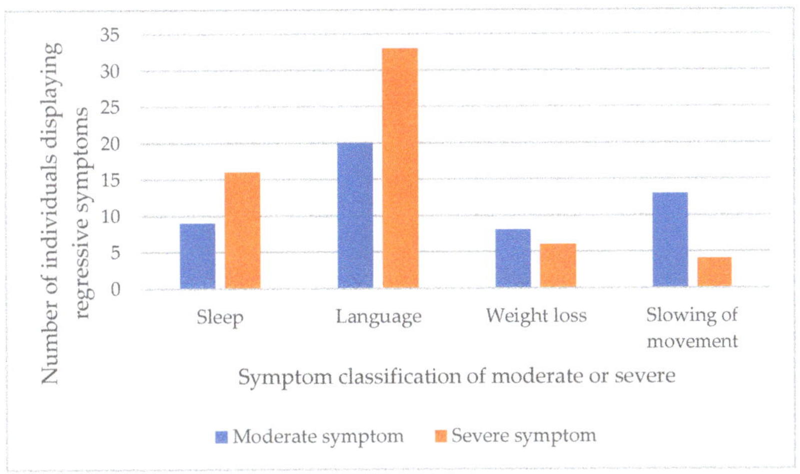

Figure 3. Number of case study patients (Group A, *n* = 39) where further details were given on the severity of the symptom.

What is striking from this analysis is the number of patients considered to have developed severe impairments of their language skills, i.e., becoming mute or losing most of their previously acquired language abilities. Sleep impairment was also more likely to be severe than moderate.

A second area of analysis sought to identify patterns of symptom co-presentation in those people with DS in the case studies (Group A). Patterns identified for this are shown in Appendix B. Sleep disorder, deterioration in language, and becoming withdrawn and disinterested were the symptoms that, if present, were associated with the full range of symptomatology, rates of co-morbidity with other symptom clusters in most cases being above 50%. In contrast, obsessive compulsive behaviours, fatigue, and abnormal blinking and gaze are associated with co-morbidity rates with other symptoms of well under 50%. What cannot be determined is whether such observations are a manifestation of where the various patients were in the time course of the regression and/or whether this is a manifestation of the maximum level of severity overall. The fact that sleep disturbance is one of the reported early symptoms perhaps indicates that the early pathophysiology of IDRS involves the hypothalamus. However, such pathophysiology would need to extend beyond the hypothalamic and limbic systems to account for the onset of motor symptoms. Mapping the course of symptom development and ultimately recovery through a longitudinal study would provide valuable information, both in terms of clinical management but also inform as to the likely underlying pathophysiological course. This analysis was not possible for the cohort studies as there was no indication as to which participants expressed multiple symptoms. As outlined in Table 3, symptoms reported include both their severe and moderate forms.

4.4. Events Preceding Regression

Records of life events occurring prior to the onset of regression were often referred to as "triggers" or "events" in the articles reviewed. The term "trigger" implies a causal effect for which we cannot be certain, in fact, it is likely that many more people with Down syndrome experience these same life events and do not develop IRDS. Among the articles reviewed this information was recorded for a total of 93 patients. Data were collected from the reviewed studies where reports of the same life event preceding a regressive episode was evident in >1 patient. Figure 4 shows the number of patients identified as experiencing such an event close to the time of their regressive episode onset.

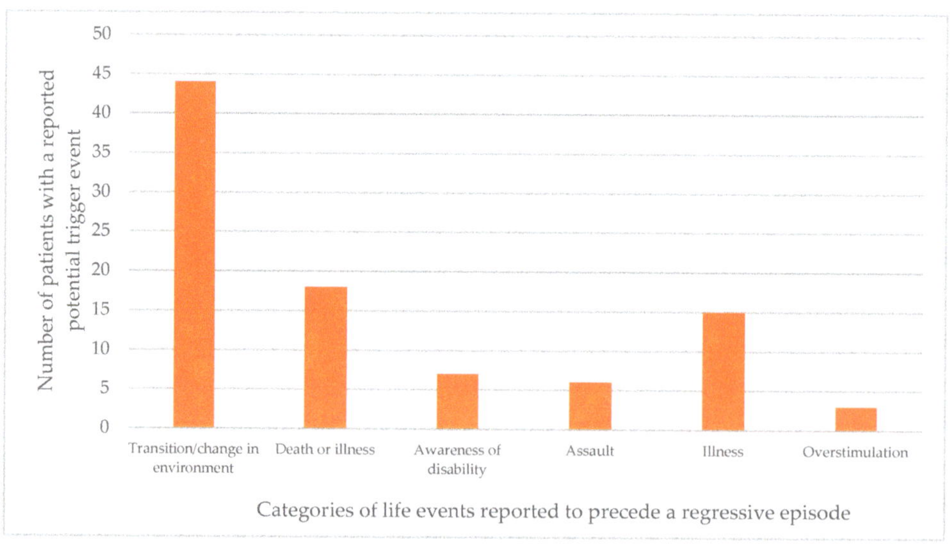

Figure 4. Number of cases with reported life events close to the time of a regressive episode (*n* = 93). Events included in this figure are those that occurred in more than one individual.

Figure 4 shows that "transition/change in environment" was the most commonly reported life event that occurred around the time of an individual regressive episode. Twelve of the 44 patients where this was suggested as a potential preceding event also included additional information regarding the circumstances. Seven cases suggested there was an association between graduation or leaving high school and the onset of a regressive period (see Figure 5).

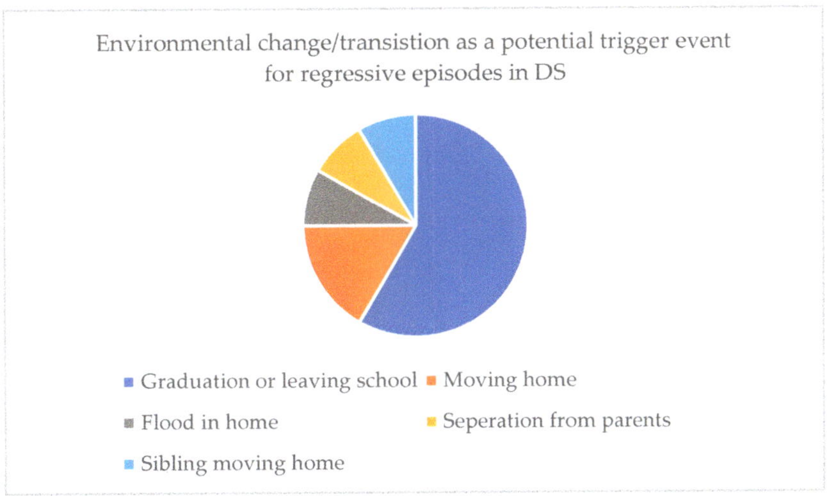

Figure 5. Breakdown of the environmental or transitional event circumstances that may have preceded a regressive episode ($n = 12$).

4.5. Brain Abnormalities

Articles that provided brain-imaging data for the participants with DS were limited. The majority reported no abnormalities using brain MRI [15,17,19] or EEGs [15,19]. MRI data in one case study showed what were reported as senile changes in the five people with DS for whom they had imaging data, including ischemic changes in the cerebral white matter, hippocampal atrophy and basal ganglia calcification [12]. Mircher et al. [18] provided the most substantial brain imaging data, with records of 15 people with DS who underwent structural MRI. Eleven were reported as having normal brain structure, while the remaining four showed indications of abnormal brain structure, specifically, thin hippocampus (1), para hippocampal sulcus verticalisation (1), cerebellar hypotrophy (1) and cortical and cerebellar hypotrophy (1). Brain abnormalities were also reported using MRI neuroimaging in a single case study of a 19-year-old man with DS, however the nature of the abnormalities were not commented on further [15]. Of 11 EEGs conducted, all patients were reported as having normal EEG activity and from 23 polysomnography tests, two patients were reported as having abnormal findings, no further details were given. Additional brain abnormalities recorded included calcification of the pallidum, pineal body and habenular commissure, as well as low signal intensity in the pallidum and high signal intensity in the pyramidal tract and crossing of the superior cerebellar peduncles. Due to the absence of a comparison group or other studies to support these findings, it is not possible to determine from these results the significance with respect to IRDS.

4.6. Medications, Interventions and Outcomes

An important aim of this review was to give insight into the types of treatments and interventions used for this condition. As there are no treatment trials or controlled trials in this area, we have compiled a list of the medications and interventions reported across all the articles and the recorded outcome for the individual with DS. These details can be seen in Appendix A. In the absence of controlled trials, it is not possible to provide a review on the efficacy of a treatment, however it is interesting to note the wide variation of medications administered, the majority of which were given for different durations, at different dosages and alongside different additional medications. Thus, it is impossible to do any direct comparisons but it is important to be aware of the variety in the current interventions used.

Treatments that were administered to more than one person with DS have been recorded in terms of the response rate that was observed, either a positive, negative or no response (see Figure 6). It should be noted that in almost none of these examples were drugs/treatments the sole intervention, there was usually a combination of treatments that is not reflected in Figure 6; dosages and treatment lengths were also not recorded. For the purpose of this analysis, each treatment and response is entered as its own entity.

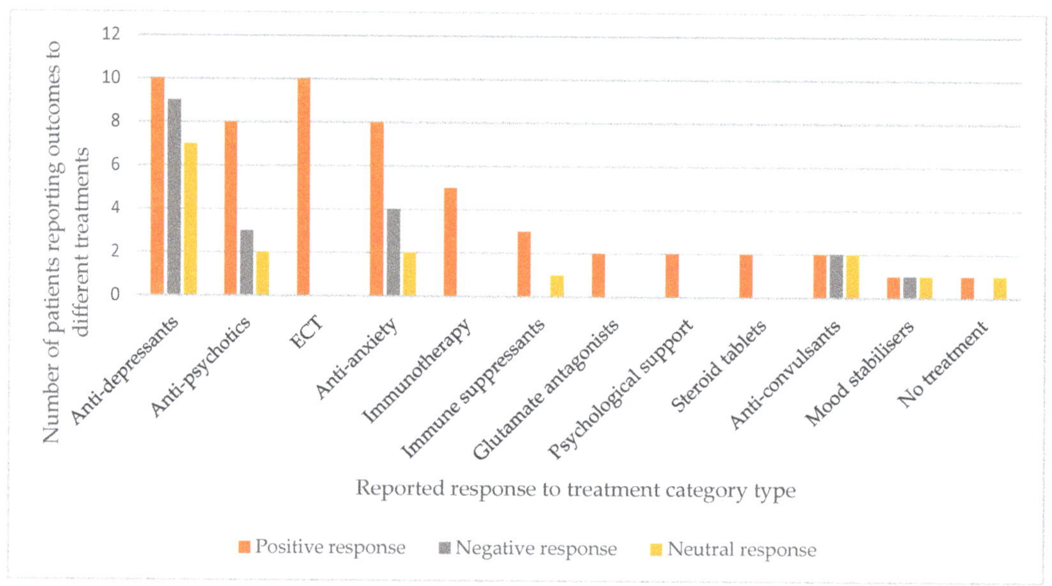

Figure 6. Bar chart showing the frequency of cases with administered drugs and treatments alongside outcomes for each type of treatment (*n* = 89). For more details on reported outcome see Appendix A.

Anti-depressants and anti-psychotics were the most frequently used medications. Anti-depressants were reported as having had almost equal positive and negative effects with slightly fewer people with DS showing no changes after the medication. Of all the anti-depressants administered, clomipramine was the only one that was reported as resulting in improvement in all four people with DS it was given to, with one complete recovery. In this case 150mg clomipramine was administered once per day for six weeks. One hundred and fifty milligrams of desipramine had been administered previously but found to be ineffective. Electro-convulsive therapy (ECT) was found to be very effective. Of the 10 patients receiving this treatment four were reported to have made a full recovery [15,18], and a further four made significant recoveries to more than 80% of their baseline functioning [20]. Other patients showed complete resolution of some behaviours [19] and a "robust response" respectively [15]. There was, however, a high occurrence of discontinuation of ECT treatment, frequently resulting in relapse. Many patients receiving ECT had a preponderance of catatonia features [15,19,20], which may explain its effectiveness. Based on the wide variety of treatments employed and the large number that exhibited either no response or a negative response, clearly there is no consensus among practitioners regarding best treatment practice and there is a great need for improvement, both in our understanding of regression in people with DS and in the potential treatments.

4.7. Prognosis

Overall, the majority of people with DS made some level of recovery from their regressive episode (Figure 7). Despite the range of different treatments used the majority appear, on the basis of clinicians' reports, to have had mostly beneficial effects. Most intriguing is the finding that all patients treated with ECT showed a positive response. However, it is noteworthy and of concern that 66% of patients made only a partial recovery and did not return to their baseline functioning. Of those making a good recovery and even those returning to near pre-regression levels it was often reported that whilst most symptoms had been resolved, a specific behaviour or symptom remained. In some of these cases the remaining symptom(s) was very detrimental to daily life, such as persistent insomnia or unresolved mutism. Only eight people with DS (20%) made a full recovery to their baseline functioning, whilst two made no improvements and a further two withdrew from their treatment program and were not followed up.

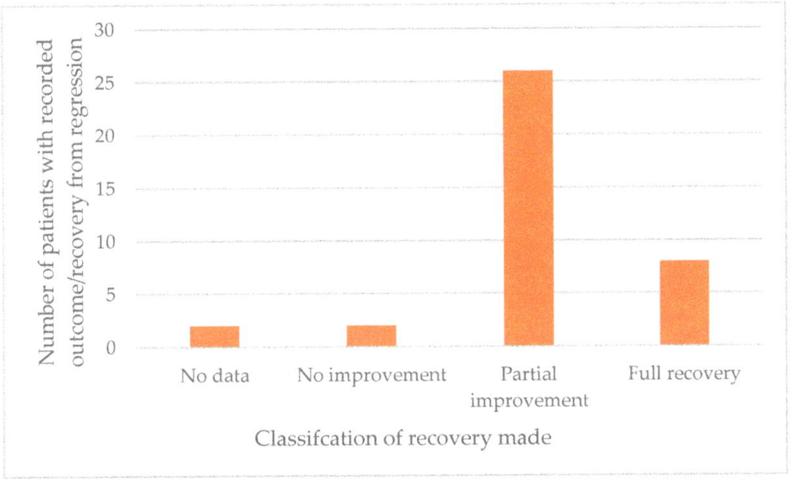

Figure 7. Recovery rate of 39 case study patients (Group A).

5. Discussion

We undertook a systematic review to increase our understanding of the nature of early regression in adults and adolescents with DS. The standalone term "regression" was used for this systematic review in order to capture as many relevant articles as possible. Based on the overwhelming number of articles referencing only a form of statistical regression this is not a method that should be recommended. Out of 1938 articles identified in the search only a small number of independent articles including observational case and cohort studies with a total of 186 people with DS were identified. Within these articles, 39 people with DS were presented as case studies/vignettes, providing qualitative details about the nature of their regression, symptoms, treatments and prognosis. Contrary to previous observations reporting a clear preponderance of females [8], males represented just under half of the population identified in this review (46.2%). Previous reports have identified the impact of IRDS [23], however, there are many differing opinions on the interpretation and classification of IRDS, including the diagnoses of reactive depressive illness [14], and catatonia [15].

We identified the use of a total of 17 different descriptive terms to label this condition within the title, abstract and keywords in 13 articles. This review is by no means the first to notice the issues surrounding the description of this condition. Worley et al. [22] initiated the use of the term DSDD in their paper detailing autistic like regression in young children with DS. Of the articles included in this review that have been published since, few have continued with the use of DSDD [8,22]. However, other papers have referenced Worley et al., and repeated the concern of the lack of a unified term [16,18].

In this review, we grouped the symptoms that were noted within the cases studies (Group A) and those seen in the cohort studies (Group B). In the absence of a population-based study an estimate of the age-specific prevalence of IRDS was not possible. However, with the case study data (Group A) the description of symptoms was detailed and qualitative in nature thus allowing further insight into the severity of the symptoms. For the purpose of this review, observations concerning the 10 most predominant symptoms from Group A were identified, and then supplemented with data from Group B, the cohort studies. There were many individual cases of very specific changes and skill loss that were not necessarily representative of others with the condition, therefore symptoms noted in less than 10% of all cases were not reported in this analysis. Our priority was to identify the symptoms that were most commonly seen, the severity of those symptoms, and whether or not they were associated with the presence of other symptoms. Determining severity was especially problematic in this review as in the absence of baseline data it was difficult to quantify skill loss. In the case studies there was some additional information given about an individual's prior behaviour and capabilities thus enabling some judgment about severity, however for consistency, the descriptive words used by the article were used to group symptoms.

In contrast to the Rosso et al. review [8], our review found that sleep disturbance was the most significant clinical feature. As in the other review, language skills decline and changes in behaviour, specifically becoming disinterested and withdrawn, were frequent. These latter symptoms were evident in around 50% of both Group A and B participants. For Group A we were able to cross-reference symptoms in an effort to establish patterns and high rates of symptom co-morbidity. We found that deterioration in sleep and speech was highly associated with almost all other symptoms, including slowness, weight loss and depression. Other associations that were seen less frequently included catatonia with skill loss, weight loss with increased slowness, and aggression with hallucinations.

Across the case studies that were reviewed there was a considerable difference in the type and amount of detail given. The majority of case studies reported in a narrative style without particular consistency in language or inclusion. This style provided the greatest detail and in-depth analysis. Studies where narratives were not used were slightly more problematic. For example, Akahoshi et al. [12] provided a table for their case reports with minimal qualitative data. It was considered that the subtleties of the symptoms may have been lost in an attempt to fit the criteria into pre-set categories. Eleven of the cases were described as showing signs of abulia (11) or hyperboulia (1), however, this symptom was not identified in any other case report narrative reviewed. Eight of these patients were also diagnosed with insomnia, the most severe of the sleep disorders reported and none with more moderate impairments. Whilst this may be a true reflection of the patients' symptoms it is also possible that a reduced number of descriptors were decided on to fit pre-determined categories. The unique reporting methods were almost more in line with that of the cohort data. A limitation we are aware of is the impact that such results can have on review data, particularly in small samples such as these.

Severity of symptoms was deduced for participants (both Groups A and B) based on the language used to describe or classify the symptoms. It was clear that within the descriptions of sleep disorders and language decline, the presence of more severe symptoms significantly outweighed the more moderate symptoms. For these categories, this equates to more patients suffering from insomnia rather than disturbed sleep, and more patients described as experiencing mutism in comparison to less severe decline in language skills. Weight loss was also of interest as equal numbers of people with DS were reported as having suffered from loss of weight/appetite or a diagnosis of anorexia nervosa. Information was not given as to the exact nature of the symptoms that led to a diagnosis of anorexia nervosa, for example the nature and extent of body dissatisfaction, and it is likely that a diagnosis of anorexia nervosa in someone with DS may be difficult to make reliably. Interestingly, in the general DS population weight loss is not often reported. More typically the opposite, over-eating and weight gain is considered problematic, therefore, relatively high numbers of people with DS experiencing weight loss and receiving the diagnosis of anorexia nervosa may be an important indicator of symptomatology that is specific to this condition. The reasons behind appetite loss and weight change were not given in the articles reviewed, however, there is a strong co-presentation of disinterest and withdrawal in those whose weight changed, including four of those who were reported to have developed anorexia nervosa. Clinical features of IRDS must be identified if there are hopes of raising awareness of this condition, Table 4 presents the profile of IRDS.

Table 4. Proposed clinical features of IRDS.

Core Symptoms and Signs	Potential Triggers for Regression	Exclusions
New onset poor sleep	Transitions	Autism spectrum disorder presents in 5 years and above
Change in language output	(e.g., changes in an individual's home/school/college routine)	Medical causes (incl. thyroid dysfunction and other conditions with autoimmune aetiology)
Abulia, withdrawal, disinterest, personality changes	Life events	New onset sensory impairment
Mood changes, loss of appetite and weight loss	Stressors	Age-related decrease in activity
Motor features–catatonia, stereotypies, extra-pyramidal signs		Other mental illness (e.g., depression)
Loss of skills (adaptive functioning)		Unlikely over the age of 40 years (dementia is possible)

On the surface, the statistics for recovery rate appear very positive. In our review only two individuals from Group A were reported as having made no improvement. However, 66% of patients made only a "partial recovery" ($n = 26$). This encompassed everything from slight improvement to near baseline functioning. Although making a recovery to "near baseline" may appear positive, in many cases one or more severe symptoms remained, including insomnia and mutism. In the case studies only eight people with DS recovered their full baseline abilities and functioning level after experiencing a regressive episode. The time course of IRDS remains uncertain in the absence of systematic longitudinal studies and it is also unknown whether there is a time point in the course of IRDS after which further recovery is unlikely.

Overall, the treatments used had varying results. Anti-depressants (including; clomipramine, bupropion, trazodone, fluvoxamine, desipramine, amitriptyline, nortriptyline and citalopram) were the most commonly administered drug type, the effects of which were almost equal between a positive, negative and no response. Similar numbers of positive and negative respondents were seen from the use of anti-anxiety drugs (mexazolam, bromazempam, benzodiazepines, lorazepam). Clomipramine exhibited a positive response in all people with DS it was given to, whilst ECT and immunotherapy appeared to have the most positive outcomes, with all patients exhibiting a positive response, although in each treatment the numbers were small (10 and 5 cases, respectively). Anti-psychotic medications (clozapine, levomepromazine, haloperidol, olanzapine, aripiprazole, ziprasidone and thiothixene) resulted in more positive responders than negative or no response. Of significant concern, both clinically and in terms of drawing any conclusions, is that in very few cases was a single treatment given independently and across the studies the length of administration and dosage varied, as were the other treatments that were alongside. For the purposes of this analysis, each intervention has been taken as an independent entity. Whilst this is far from ideal, it is impossible to accurately record response to a single treatment when multiple are prescribed in conjunction. There were no controlled treatment trials among the studies reviewed and in the absence of such trials it is impossible to be certain whether or not reported improvements following treatment are a manifestation of a treatment effect or just an indication of the natural history of IRDS in that individual.

Many of the symptoms presenting in IRDS are also seen in other conditions, such as depression and anxiety, or are seen in response to stress. These conditions may present atypically in people with DS [13], including the loss of functional skills, sleep impairment and reduced language [24]. In several studies, it was noted that people with DS were not experiencing depression prior to the episode, and many individuals were unresponsive to psychotropic drugs. Furthermore, some of the behaviours and changes, such as mutism, were never reversed despite making an otherwise complete recovery. This extent of persistent loss is not commonly seen in pure mood disorders.

A striking feature of IRDS is the age of onset, which is at an age when brain development is in its final stages and still susceptible to being disturbed. This age of risk for IRDS is later than that of classical autistic regression, or that observed in other rare neurodevelopmental disorders, and earlier than would be expected for dementia. Furthermore, the majority of studies have not reported abnormal imaging (MRI) or encephalographic (EEG) findings [15,17,19]. Other data showed senile changes in the five people with DS that received an MRI scan; however, there was no comparison or longitudinal data available [12]. Although routine medical screening (such as for thyroid disease, coeliac disease and vitamin D deficiency) undertaken in people with DS presenting with possible regression may be abnormal, it is unlikely such abnormalities are causative [4]. However, treatment of co-occurring medical conditions may improve prognosis.

The presentation of IRDS has been likened to that of dementia. Although the age of onset of dementia in people with DS is early compared to the general population it is still considerably later than the typical age of onset of IRDS. Myers and Pueschel [10] included a case study of a 44-year-old person with DS, excluded from this review due to our age inclusion criteria, who exhibited very similar symptoms to that of the younger participant, but the psychiatric diagnosis was AD. The major differences between the type of regression discussed in this review and dementia is the age of onset and, most crucially, that patients often show some recover from IRDS. No recovery is seen in those diagnosed with AD.

Many other diagnoses have been considered to explain this early regression, including psychotic illness, catatonia, mania, depression and anxiety. Autoimmune encephalopathy and mitochondrial dysfunction have been considered as possible underlying mechanisms [25,26]. Immune abnormalities are typical in children with DS [27]) and may be of aetiological significance. Interestingly one of the most effective treatments reported in this review were drugs impacting on the immune system, where all but one individual (who discontinued their use due to negative side effects) saw positive results, including one full recovery, and full recovery with the exception of persistent insomnia.

At present the underlying cause(s) of IRDS and why it appears to be specific to people with DS are unknown. One key question is whether we know enough about IRDS at present to argue that IRDS is a specific condition with a common, but as yet unknown, aetiology. Rosso et al. [8] argue that, as the cause(s) of IRDS is unknown, a full clinical work up in all cases of young people with DS who report such changes is required to identify possible explanatory and potentially different diagnoses. If, on the other hand, all clinically diagnosed cases of IRDS are in fact considered to have a common cause a further question is whether that is best explained as a consequence of the atypical presentation of a known co-morbid condition (e.g., catatonia), which occurs in other populations but happens to be more common in people with DS; or whether IRDS is a condition that has an aetiology that is unique and specific to DS, and ultimately can be linked back to the presence of trisomy 21. The paper by Miles et al. [20] interestingly reported on seven patients with DS who had developed regression and the authors proposed that in each case this regression was due to catatonia. Using the Bush–Francis Catatonia Rating Scale and demonstrating a positive response to intravenous lorazepam they argued that this was true catatonia. They also reported a good response to ECT and other treatments although recovery was often not maintained. Such observations would suggest that regression is due to a condition (in this case catatonia) that can affect anyone, but that people with DS at a specific age are particularly vulnerable. However, although people with other neurodevelopmental syndromes may have syndrome specific neuropsychiatric risks, to our knowledge similar episodes of regression in the same age range are rare in people with other neurodevelopmental syndromes. This is more in support of the hypothesis that IRDS is a DS specific condition. An exception to this is cases of regression in people with SHANK3 mutations (Phelan–McDermid syndrome), reported in two papers [28,29]. For people with Prader-Willi Syndrome, particularly those with the maternal uniparental disomy form, a regressive type of clinical deterioration can be seen in the same age group as IRDS but the symptomatology is much more obviously that of a psychotic illness [30]. Although there are some differences in symptom prevalence [8] the symptomatology that is described in both reviews would appear to separate IRDS from other potential aetiologies, such as affective disorder. However, such uncertainties as those described above are clearly hindering a rational approach to treatment development.

It is unknown whether there are specific risk markers and biomarkers that are either an indicator of vulnerability to IRDS or can map to the presence and severity of the IRDS and are potential indicators for underlying causative mechanisms. These need to be studied and markers identified and then followed over time. Whether there is a relationship between the risk for regression and the later risk of dementia is unknown and specifically whether there are genetic markers that affect the vulnerability to dementia (e.g., ApoE genotype and soluble TREM-2) are also associated with regression. And whether biomarkers of other potential mechanisms (e.g., myelodysplasia, leukopenia, macrocytosis), and inflammatory markers (e.g., inflammation related factors, cytokines) are important and need further investigation. Many of these markers have been identified as being associated with major psychiatric disorders.

A striking observation is the potential role of transitional life events and changes in environment as a potential trigger for regression. IRDS may be best considered as a condition triggered by stress and occurring in people with DS who have some additional genetic or acquired vulnerability (low resilience). Alternatively, IRDS is due to the occurrence of an acquired condition that results in a direct, and initially adaptive response in the brain to the insult and subsequently in a temporary and adverse effect on brain function. For example, an acquired insult may lead to the development of an inflammatory response in the brain resulting in an encephalopathy that subsequently completely or partially resolves.

Limitations

There are several limitations with this review article that have mostly been discussed in the course of this review. First, the search strategy used to capture relevant articles vastly over-included studies and returned a huge number of results that were irrelevant. The difficulty with this search centred on the labelling words used to describe this condition, and that there is not a unified term. Our compromise, after trialling many search strategies, was to include the word "regression" without additional restrictions. While this returned the best results in so far as the few relevant articles, there were also a plethora of studies referencing only statistical regression that needed to be manually eliminated. It is recognised that it is not good practice to have such a large difference between the number of articles sourced and those retained, it was deemed necessary based on the confusion seen in the terminology used for this condition. Although the methodology was time consuming and over-inclusive, using the particular search criteria selected we are confident that the correct papers have been identified and included.

The articles included in this paper were all observational studies, both case and cohort studies. The positive aspects of observational data are that records are usually qualitative and highly descriptive, allowing for cross-referencing between symptoms to be explored and person-by-person outcomes evaluated based on interventions. The negatives are that the quality of evidence is low. Cochrane's levels of evidence quality [31] describe randomised controlled trials as giving the highest quality of evidence and observational studies amongst the lowest. As such any conclusions drawn from the studies must be very cautiously considered. It is our hope that further insight and raising awareness will lead to a greater interest in research of this condition and promote controlled trials in the future.

Despite these limitations this review has provided insight into an under-researched condition with significant impacts on people with DS and their families. This review intends to bring to light a serious condition affecting a minority of young adults and adolescents, many of whom never recover their baseline functioning. It is important that we now seek to focus on prevention and treatment of this condition.

Author Contributions: Conceptualisation M.W., A.H. and S.Z.; methodology, M.W., software M.W.; validation M.W., A.H. and S.Z., formal analysis, M.W., investigation, M.W., resources, M.W., writing—original draft, M.W., A.H. and S.Z. All authors have read and agreed to the published version of the manuscript.

Funding: This research was funded by the UK Down Syndrome Association.

Institutional Review Board Statement: Not applicable.

Informed Consent Statement: Not applicable.

Data Availability Statement: Not applicable.

Acknowledgments: The authors would like to recognize the support from the UK Down Syndrome Association.

Conflicts of Interest: The authors declare no conflict of interest.

Appendix A

Table A1. Table showing the range of treatments administered across all papers on an individual basis alongside reported outcome.

Treatment/Intervention	Administered *n*	Positive Response	Negative Response	No Response
SSRI—fluvoxamine	5	3 "Symptoms improved" "Moderate improvement" "Significantly better"	0	2 "No improvement" "No change"
Amantadine	2	2 "Partial improvement" "Partial improvement"		
Levomepromazine	2	2 "Partial improvement" "Partial improvement"		
Haloperidol	4	3 "Improvement" "Partially improved" "Improvement for 6 months"		1 "Unsuccessful"
Mexazolam	1	1 "Improvement"		
Bromazepam	1	1 "Partial improvement"		
Carbamazepine	2	1 "Partial improvement"		1 "Unsuccessful"
Clomipramine	4	3 "Partial improvement" "Partial improvement" "Complete recovery" "Slight improvement, side effects"		

Table A1. *Cont.*

Treatment/Intervention	Administered *n*	Positive Response	Negative Response	No Response
Romethazine	1	1 "Partial improvement"		
Lorazepam	10	4 "Partial improvement" "80% return to baseline" "Responded to" "Showed increase" "Significant improvement"	1 "Negative effects"	5 "No consistent improvement" "No consistent improvement" "No consistent improvement" "No change" "Ineffective"
Methylprednisolon	3	3 "Dramatic improvement" "Immediate improvement" "Many behaviours resolved"		
IVIG	4	4 "Full recovery" "Steady improvement" "Lots of symptoms resolved" "Resolution of everything except insomnia"		
Mycephenolate	1			1 Discontinued no data
Oral steroid	2	2 "Immediate improvement" "Lots of resolved symptoms"		
Rituximab	1	1 "Improvement"		
Electro-convulsive therapy	10	10 "Some behaviours completely resolved" "Complete recovery" "Complete recovery" "Complete recovery" "Complete recovery" "Robust response" "Significant improvement" "Return to almost baseline" "Excellent response" Strong response"		
Benzodiazepines	1	1 "Partial improvement"		
Anti depressants	1	1 "Steady return to baseline"		
Positive airway pressure for obstructive sleep apnoea	1	1 "Steady return to baseline"		
Psychological support	2	2 "Steady return to baseline" "Moderate improvement for 6 months"		
Donepezil	1	1 "Return to baseline"		
Acetylcholinesterase inhibitor	1	1 "Return to baseline"		
Bupropion	1		1 "Worsening of catatonia and further decline"	
Trazodone	4	1 "Some improvement"	2 "Worsening of catatonia and further decline" "Worsening and decline"	1 "Unsuccessful"

Table A1. Cont.

Treatment/Intervention	Administered *n*	Positive Response	Negative Response	No Response
Olanzapine	1		1 "Worsening of catatonia and further decline"	
Aripiprazole	1		1 "Worsening of catatonia and further decline"	
Ziprasidone	1		1 "Worsening of catatonia and further decline"	
Lithium	3	1 "Improvement"	1 "Worsening of catatonia and further decline"	1 "No improvement"
Clozapine	1	1 "85% return to baseline"		
Desipramine	6	1 "Moderate improvement for 6 months"	1 "Worsening of symptoms"	4 "No change" "No effect" "Unsuccessful" "Unsuccessful"
Thiothixine	2	1 "Complete recovery"		1 "No change"
Amitriptyline	1	1 "Complete recovery"		
Nortriptyline	1			1 "Unsuccessful"
Clonazepam	2	1 "Partial improvement"		1 "Unsuccessful"
Ethosuximide	1			1 "No effect"
Lidexamfetamine	1		1 "Worsening"	
SSRI citalopram	1		1 "Worsening"	
Amiloride	1	1 "Improvement"		
Lamotrigine	1			1 "No effect"
Antipsychotic treatment	1	1 "Good response"		

Note: dosages and additional medications/treatments are not reported.

Appendix B

Table A2. Table showing co-presentation of symptoms in Group A patients.

	Sleep (n = 22)	%	Language (n = 14)	%	Withdrawal and disinterest (n = 14)	%	Slowness and immobility (n = 14)	%	Weight loss and anorexia (n = 12)	%	Depression (n = 11)	%	Hallucinations (n = 11)	%	Abulia (n = 10)	%	Skill loss (n = 10)	%	Catatonia (n = 8)	%	Aggression (n = 7)	%	Irritability (n = 6)	%	Obsessive compulsions (n = 6)	%	Fatigue (n = 5)	%	Abnormal blinking and gaze (n = 4)	%
Sleep	0		10	71	9	64	8	57	10	83	9	82	7	64	7	70	6	60	6	75	6	86	2	33	5	83	4	80	2	50
Language	10	45	0		3	21	7	50	4	33	7	64	6	55	2	20	6	60	6	75	4	57	2	33	1	17	0	0	2	50
Withdrawal and disinterest	9	41	3	21	0		6	43	5	42	5	45	4	36	8	80	2	20	0	0	2	29	3	50	4	67	3	60	0	0
Slowness and immobility	8	36	7	50	6	43	0		7	58	5	45	4	36	2	20	4	40	5	63	2	29	1	17	3	50	2	40	4	100
Weight loss and anorexia	10	45	4	29	5	36	7	50	0		3	3	4	36	2	20	2	20	2	25	2	29	1	17	1	17	4	80	2	50
Depression	9	41	7	50	5	36	5	36	3	25	0		4	36	0	0	2	20	6	75	2	29	1	17	0	0	0	0	2	50
Hallucinations	7	32	6	43	4	29	4	29	4	33	4	36	0		0	0	4	40	0	0	2	29	0	0	0	0	0	0	3	75
Abulia	7	32	2	14	8	57	2	14	2	17	4	36	0	0	0		0	0	2	25	2	29	0	0	0	0	0	0	3	75
Skill loss	6	27	6	43	2	14	4	29	2	17	2	18	6	55	2	20	0		2	25	0	0	1	17	0	0	2	40	0	0
Catatonia	6	27	6	43	0	0	5	36	2	17	2	18	0	0	2	20	6	60	0		0	0	0	0	0	0	3	60	0	0
Aggression	6	27	4	29	2	14	2	14	2	17	2	18	2	18	2	20	2	20	0	0	0		2	33	0	0	1	20	0	0
Irritability	2	9	2	14	3	21	1	7	1	8	1	9	0	0	3	30	1	10	2	25	2	29	0		2	33	0	0	2	50
Obsessive compulsions	5	23	1	7	4	29	3	21	1	8	0	0	0	0	0	0	4	40	0	0	0	0	4	57	0		0	0	0	0
Fatigue	4	18	0	0	3	21	2	14	4	33	0	0	0	0	0	0	0	0	0	0	2	29	2	33	0	0	0		1	25
Abnormal blinking and gaze	2	9	2	14	0	0	4	29	2	17	2	18	3	27	1	10	2	20	1	13	0	0	0	0	0	0	1	20	0	

References

1. Antonakaris, S.E.; Stotko, B.G.; Rafii, M.S.; Strydom, A.; Pape, S.E.; Bianchi, D.W.; Sherman, S.L.; Reeves, R.H. Down Syndrome. *Nat. Rev. Dis. Primers* **2020**, *6*, 1–20. [CrossRef]
2. Rasmussen, S.A.; Whitehead, N.; Collier, S.A.; Frías, J.L. Setting a public health research agenda for Down syndrome: Summary of a meeting sponsored by the centers for disease control and prevention and the national down syndrome society. *Am. J. Med. Genet. Part. A* **2008**, *146*, 2998–3010. [CrossRef] [PubMed]
3. Dykens, E.M.; Shah, B.; Davis, B.; Baker, C.; Fife, T.; Fitzpatrick, J. Psychiatric disorders in adolescents and young adults with Down syndrome and other intellectual disabilities. *J. Neurodev. Disord.* **2015**, *7*, 9. [CrossRef]
4. Santoro, S.L.; Cannon, S.; Capone, G.; Franklin, C.; Hart, S.J.; Hobensack, V.; Kishnani, P.S.; Macklin, E.A.; Manickam, K.; McCormick, A.; et al. Unexplained regression in Down syndrome: 35 cases from an international Down syndrome database. *Genet. Med.* **2019**, *22*, 767–776. [CrossRef] [PubMed]
5. Mann, D.M.; Esiri, M.M. The pattern of acquisition of plaques and tangles in the brains of patients under 50 years of age with Down's syndrome. *J. Neurol. Sci.* **1989**, *89*, 169–179. [CrossRef]
6. Annus, T.; Wilson, L.R.; Hong, Y.T.; Acosta-Cabronero, J.; Fryer, T.D.; Cardenas-Blanco, A.; Smith, R.; Boros, I.; Coles, J.P.; Aigbirhio, F.I.; et al. The pattern of amyloid accumulation in the brains of adults with Down syndrome. *Alzheimer's Dement.* **2016**, *12*, 538–545. [CrossRef]
7. Hardy, J.; Selkoe, D.J. The amyloid hypothesis of Alzheimer's disease: Progress and problems on the road to therapeutics. *Science* **2012**, *297*, 353–356. [CrossRef]
8. Rosso, M.; Fremion, E.; Santoro, S.L.; Oreskovic, N.M.; Chitnis, T.; Skotko, B.G.; Santoro, J.D. Down Syndrome Disintegrative Disorder: A Clinical Regression Syndrome of Increasing Importance. *Pediatrics* **2020**, *145*, e20192939. [CrossRef]
9. Moher, D.; Liberati, A.; Tetzlaff, J.; Altman, D.G.; Altman, D.; Antes, G.; Atkins, D.; Barbour, V.; Barrowman, N.; Berlin, J.A.; et al. Preferred reporting items for systematic reviews and meta-analyses: The PRISMA statement. *J. Chin. Integr. Med.* **2009**, *7*, 889–896. [CrossRef]
10. Myers, B.A.; Pueschel, S.M. Major depression in a small group of adults with down syndrome. *Res. Dev. Disabil.* **1995**, *16*, 285–299. [CrossRef]
11. Capone, G.T.; Aidikoff, J.M.; Goyal, P. Adolescents and Young Adults With Down Syndrome Presenting to a Medical Clinic With Depression: Phenomenology and Characterization Using the Reiss Scales and Aberrant Behavior Checklist. *J. Ment. Health Res. Intellect. Disabil.* **2011**, *4*, 244–264. [CrossRef]
12. Akahoshi, K.; Matsuda, H.; Funahashi, M.; Hanaoka, T.; Suzuki, Y. Acute neuropsychiatric disorders in adolescents and young adults with down syndrome: Japanese case reports. *Neuropsychiatr. Dis. Treat.* **2012**, *8*, 339–345. [CrossRef]
13. Stein, D.S.; Munir, K.M.; Karweck, A.J.; Davidson, E.J.; Stein, M.T. Developmental regression, depression, and psychosocial stress in an adolescent with Down syndrome. *J. Dev. Behav. Pediatr.* **2013**, *34*, 216–218. [CrossRef] [PubMed]
14. Capone, G.T.; Aidikoff, J.M.; Taylor, K.; Rykiel, N. Adolescents and young adults with down syndrome presenting to a medical clinic with depression: Co-morbid obstructive sleep apnea. *Am. J. Med. Genet. Part A* **2013**, *161*, 2188–2196. [CrossRef] [PubMed]
15. Ghaziuddin, N.; Nassiri, A.; Miles, J.H. Catatonia in down syndrome; a treatable cause of regression. *Neuropsychiatr. Dis. Treat.* **2015**, *11*, 941–949. [CrossRef]
16. Jacobs, J.; Schwartz, A.; McDougle, C.J.; Skotko, B.G. Rapid clinical deterioration in an individual with Down syndrome. *Am. J. Med. Genet. Part. A* **2016**, *170*, 1899–1902. [CrossRef] [PubMed]
17. Tamasaki, A.; Saito, Y.; Ueda, R.; Ohno, K.; Yokoyama, K.; Satake, T.; Sakuma, H.; Takahashi, Y.; Kondoh, T.; Maegaki, Y. Effects of donepezil and serotonin reuptake inhibitor on acute regression during adolescence in Down syndrome. *Brain Dev.* **2016**, *38*, 113–117. [CrossRef]
18. Mircher, C.; Cieuta-Walti, C.; Marey, I.; Rebillat, A.-S.; Cretu, L.; Milenko, E.; Conte, M.; Sturtz, F.; Rethore, M.-O.; Ravel, A. Acute Regression in Young People with Down Syndrome. *Brain Sci.* **2017**, *7*, 57. [CrossRef]
19. Cardinale, K.M.; Bocharnikov, A.; Hart, S.J.; Baker, J.A.; Eckstein, C.; Jasien, J.M.; Gallentine, W.; Worley, G.; Kishnani, P.S.; Van Mater, H. Immunotherapy in selected patients with Down syndrome disintegrative disorder. *Dev. Med. Child. Neurol.* **2018**, *61*, 847–851. [CrossRef] [PubMed]
20. Miles, J.H.; Takahashi, N.; Muckerman, J.; Nowell, K.P.; Ithman, M. Catatonia in down syndrome: Systematic approach to diagnosis, treatment and outcome assessment based on a case series of seven patients. *Neuropsychiatr. Dis. Treat.* **2019**, *15*, 2723–2741. [CrossRef]
21. Castillo, H.; Patterson, B.; Hickey, F.; Kinsman, A.; Howard, J.M.; Mitchell, T.; Molloy, C.A. Difference in age at regression in children with autism with and without Down syndrome. *J. Dev. Behav. Pediatr.* **2008**, *29*, 89–93. [CrossRef]
22. Worley, G.; Crissman, B.G.; Cadogan, E.; Milleson, C.; Adkins, D.W.; Kishnani, P.S. Down syndrome disintegrative disorder: New-onset autistic regression, dementia, and insomnia in older children and adolescents with down syndrome. *J. Child. Neurol.* **2015**, *30*, 1147–1152. [CrossRef]
23. Prasher, V. Disintegrative syndrome in young adults. *Int. J. Psychiatr. Med.* **2002**, *19*, 101–102. [CrossRef] [PubMed]
24. Cooper, S.-A.; Collacott, R.A. Clinical Features and Diagnostic Criteria of Depression in Down Syndrome. *Br. J. Psychiatry* **1994**, *165*, 399–403. [CrossRef] [PubMed]
25. Brodtmann, A. Hashimoto encephalopathy and down syndrome. *Arch. Neurol.* **2009**, *66*, 663–666. [CrossRef] [PubMed]

26. Mattman, A.; Jarvis-Selinger, S.; Mezei, M.M.; Salvarinova-Zivkovic, R.; Alfadhel, M.; Lillquist, Y. Mitochondrial disease clinical manifestations: An overview. *Br. Columbia Med. J.* **2011**, *53*, 183–187.
27. Kusters, M.A.A.; Verstegen, R.H.J.; Gemen, E.F.A.; De Vries, E. Intrinsic defect of the immune system in children with Down syndrome: A review. *Clin. Exp. Immunol.* **2009**, *156*, 189–193. [CrossRef]
28. Kohlenberg, T.M.; Trelles, M.P.; McLarney, B.; Betancur, C.; Thurm, A.; Kolevzon, A. Psychiatric illness and regression in individuals with Phelan-McDermid syndrome. *J. Neurodev. Disord.* **2020**, *12*, 7. [CrossRef]
29. Bey, A.L.; Gorman, M.P.; Gallentine, W.; Kohlenberg, T.M.; Frankovich, J.; Jiang, Y.-H.; Van Haren, K. Subacute Neuropsychiatric Syndrome in Girls With SHANK3 Mutations Responds to Immunomodulation. *Pediatrics* **2020**, *145*, e20191490. [CrossRef]
30. Soni, S.; Whittington, J.; Holland, A.J.; Webb, T.; Maina, E.N.; Boer, H.; Clarke, D. The phenomenology and diagnosis of psychiatric illness in people with Prader–Willi syndrome. *Psychol. Med.* **2008**, *38*, 1505–1514. [CrossRef] [PubMed]
31. Higgins, J.P.T.; Altman, D.G.; Gøtzsche, P.C.; Jüni, P.; Moher, D.; Oxman, A.D.; Savović, J.; Schulz, K.F.; Weeks, L.; Sterne, J.A.C. The Cochrane Collaboration's tool for assessing risk of bias in randomised trials. *BMJ* **2011**, *343*, 1–9. [CrossRef] [PubMed]

Article

Selective Serotonin Reuptake Inhibitors for the Treatment of Depression in Adults with Down Syndrome: A Preliminary Retrospective Chart Review Study

Robyn P. Thom [1,2,3,†], Michelle L. Palumbo [1,2,4,†], Claire Thompson [1], Christopher J. McDougle [1,2,3,*] and Caitlin T. Ravichandran [1,2,3,5]

1   Lurie Center for Autism, 1 Maguire Road, Lexington, MA 02421, USA; rthom@mgh.harvard.edu (R.P.T.); mlpalumbo@partners.org (M.L.P.); cthompson44@mgh.harvard.edu (C.T.); cravichandran@mclean.harvard.edu (C.T.R.)

2   Massachusetts General Hospital, 55 Fruit Street, Boston, MA 02114, USA

3   Department of Psychiatry, Harvard Medical School, 25 Shattuck Street, Boston, MA 02115, USA

4   Department of Pediatrics, Harvard Medical School, 25 Shattuck Street, Boston, MA 02115, USA

5   McLean Hospital, 115 Mill Street, Belmont, MA 02478, USA

*   Correspondence: cmcdougle@mgh.harvard.edu; Tel.: +1-781-860-1700 or +1-781-860-1766

†   Equal contributions.

Citation: Thom, R.P.; Palumbo, M.L.; Thompson, C.; McDougle, C.J.; Ravichandran, C.T. Selective Serotonin Reuptake Inhibitors for the Treatment of Depression in Adults with Down Syndrome: A Preliminary Retrospective Chart Review Study. *Brain Sci.* **2021**, *11*, 1216. https://doi.org/10.3390/brainsci11091216

Academic Editor: Corrado Romano

Received: 16 July 2021
Accepted: 13 September 2021
Published: 15 September 2021

Publisher's Note: MDPI stays neutral with regard to jurisdictional claims in published maps and institutional affiliations.

Abstract: Background: Depression is a common psychiatric comorbidity in individuals with Down syndrome (DS), particularly adults, with an estimated lifetime prevalence of at least 10%. The current literature on the treatment of depression in adults with DS is limited to case series published more than two decades ago, prior to the widespread use of modern antidepressant medications such as selective serotonin reuptake inhibitors (SSRIs). The purpose of this retrospective chart review study was to examine the effectiveness, tolerability, and safety of SSRIs for depression in adults with DS. Methods: Medical records of 11 adults with DS and depression were reviewed. Assignment of scores for severity (S) of symptoms of depression and improvement (I) of symptoms with treatment with an SSRI was made retrospectively using the Clinical Global Impression Scale (CGI). Demographic and clinical characteristics of the study population, SSRI name, dose, and duration of treatment; and adverse effects were also recorded. Results: All 11 patients (7 male, 4 female; mean age = 27.2 years, range 18–46 years) completed a 12-week treatment course with an SSRI. The median duration of time after initiation of the SSRI covered by record review was 2.1 years, with a range of 24 weeks to 6.7 years. Nine of the 11 patients (82%; 95% CI 52%, 95%) were judged responders to SSRIs based on a rating of "much improved" or "very much improved" on the CGI-I after 12 weeks of treatment (median time of follow-up was 14.4 weeks, with a range of 12.0–33.0 weeks). Adverse effects occurred in four patients (36%). The most common adverse effects were daytime sedation and anger. Conclusions: In this preliminary retrospective study, the majority of patients responded to a 12-week course of SSRI treatment and some tolerated long-term use. Controlled studies are needed to further assess the efficacy, tolerability, and safety of SSRIs for the treatment of depression in adults with DS.

Keywords: Down syndrome; depression; selective serotonin reuptake inhibitor

1. Background

Down syndrome (DS) or trisomy 21, is a common genetic syndrome, resulting from an extra copy of chromosome 21. According to the Centers for Disease Control and Prevention, about 6000 babies are born with DS each year in the United States [1], and overall, DS occurs in about 1 in 700 live births [2]. Down syndrome commonly includes characteristic physical features, a variable degree of cognitive impairment, and several medical comorbidities. Medical comorbidities commonly associated with DS include congenital heart defects, thyroid disease, gastrointestinal problems, hematological disorders, hearing loss, ocular disorders, and obstructive sleep apnea [3]. In addition to elevated rates of medical comorbidity, individuals with DS have an increased risk of psychiatric disorders compared to the general population [4–6]. Prevalence rates of psychiatric comorbidities have been reported to be as high as 38% and 35% in children and adults with DS, respectively [4,5]. Externalizing symptoms, such as oppositionality, impulsivity, and hyperactivity, are more common in children with DS, whereas internalizing symptoms, such as depression, anxiety, and social avoidance, become more prevalent in adolescence and adulthood [7–9].

Depression is a common comorbidity in adults with DS, with reported prevalence rates ranging from 6–18% [9–13]. A recent study which included 605 individuals with DS from England and Wales demonstrated that 12.4% of younger adults (16–35 years) and 18.4% of older adults ($\geq$36 years) had a history of depression based on medical record review [13]. Females and males with DS had a four- and five-fold increased risk of depression, respectively, compared to the general United Kingdom adult population [13]. In a separate longitudinal cohort study, 134 adolescents and adults with DS ($\geq$16 years) participated in a detailed psychiatric assessment with psychiatrists who had expertise in DS at baseline and two years later [12]. The two-year incidence of a major depressive episode in this study was 5.2% [12]. Adults with DS have several unique risk factors for developing depression compared to the general population, including cognitive impairment [14], reduced serotonin brain tissue concentration in post-mortem studies [15], high prevalence of thyroid disorders [3], and significant emotional stressors related to the transition to adulthood and loss of school-based programming and services.

Depression is often underrecognized and undertreated in adults with DS. There are several diagnostic challenges related to the inherent communicative and cognitive limitations. While the clinical characteristics of depression in individuals with DS are often similar to those seen in the general population, including sad mood, anhedonia, decreased appetite and weight loss, social withdrawal, reduced speech, low energy, and psychomotor slowing [6], individuals with DS may have difficulty expressing depressive cognitions such as guilt, worthlessness, self-deprecation, or thoughts of suicide [10]. These clinical features may necessitate taking into account behavioral observations and caregiver reports rather than strict application of diagnostic criteria. A retrospective study assessing the clinical features of depression in DS reported that when strict Diagnostic and Statistical Manual of Mental Disorders, Third Edition (DSM-III-R) criteria were applied, only 50% of depressive episodes diagnosed by expert clinicians met the full criteria [10]. This study also demonstrated that depression was frequently misdiagnosed as dementia in individuals with DS and therefore left untreated [10]. In a retrospective study of 42 adults with DS, not all patients with depression received pharmacotherapy, and no patients received a second medication trial if the first was ineffective [10], suggesting undertreatment of a generally treatable psychiatric comorbidity.

Data on effective treatment approaches for depression in DS are lacking, and systematic studies on the treatment of depression have been highlighted as a critically needed area of research [16]. The published literature on pharmacotherapy for depression in DS is limited to case reports and case series [6]. No systematic studies on the effectiveness, tolerability, and safety of antidepressants in DS have been published. Three case series were published more than 20 years ago, prior to the widespread use of modern antidepressants, reporting the clinical response to tricyclic antidepressants (TCAs) [17–19] and one of these case series also reported on three patients' response to fluoxetine, a selective serotonin reuptake inhibitor (SSRI) [18]. All three patients who received treatment with fluoxetine had a positive response, two of whom had previously failed to respond to TCAs. Of the remaining six patients described in this case series, three other patients responded to TCAs, one responded to a first-generation antipsychotic, and two did not receive pharmacologic treatments. Medication side effects are not reported in any of these case series.

Selective serotonin reuptake inhibitors are a class of medications which include fluoxetine, paroxetine, sertraline, fluvoxamine, citalopram, and escitalopram. They selectively block the uptake of serotonin and have several Food and Drug Administration (FDA) indications, including for the treatment of depression, anxiety disorders, obsessive-compulsive disorder, and posttraumatic stress disorder. Selective serotonin reuptake inhibitors have largely replaced the TCAs as the first-line treatment for depressive disorders due to similar efficacy, improved tolerability, and a much safer side effect profile [20]. Unlike TCAs, SSRIs are generally nonlethal in overdose, are not associated with cardiac toxicity, and do not lower the seizure threshold. Modern clinical practice guidelines include SSRIs among the first-line medications for the treatment of depression [21]. Tricyclic antidepressants are considered second-line medications, only to be used after the failure of one or more first-line medications [21]. In clinical practice, SSRIs are typically the first class of medications used to treat depression. A recent study demonstrated that SSRIs comprised 93% of first-line medications for depression in primary care [22]. Selective serotonin reuptake inhibitors have a relatively benign side effect profile and are generally well tolerated. The most common side effects include impaired sexual functioning, sleepiness, and weight gain; 25% of patients consider side effects to be either "very bothersome" or "extremely bothersome" [23]. Despite the widespread use of SSRIs in the general population, their use in patients with DS has only been reported in three patients in a single case series [18].

This study aims to provide preliminary naturalistic data on whether treatment with SSRIs is effective, tolerable, and safe in reducing depression symptoms in adults with DS.

2. Methods

2.1. Study Participants

Patients potentially eligible for inclusion were identified using the research patient registry for a large hospital network in the Northeastern United States. The medical records of patients with DS, newly initiated on an SSRI (fluoxetine, paroxetine, sertraline, fluvoxamine, citalopram, or escitalopram) at age 18 years or older for the treatment of a depressive disorder (major depressive disorder, dysthymia, persistent depressive disorder, or mood disorder not otherwise specified), who had at least one follow-up visit after SSRI initiation, treated by a psychiatrist at a tertiary care center outpatient neurodevelopmental disorders clinic in the Northeastern United States from 2011–2021 were identified for detailed review. Patients who did not meet these criteria were not included in the study. Routinely collected clinical notes were retrospectively reviewed, extracted, and coded into a RedCap database. The study was approved as an exempt study by the local institutional review board. The depressive disorder diagnosis was made during the course of clinical care by board-certified psychiatrists (MLP and CJM) with expertise in treating adults with developmental disabilities. The diagnosis was corroborated by review of clinical documentation by a second board-certified psychiatrist (RPT).

2.2. Outcomes

Data were collected retrospectively from the medical record. Demographic data, diagnostic information, including severity of intellectual disability, language ability, medical, and psychiatric comorbidities, concomitant medication and non-medication treatments, and symptom changes were collected. If standardized cognitive testing was not available, intellectual ability was clinically determined. Details of the SSRI trial, including medication name, duration of treatment, starting and maximum dosage, and adverse effects, were also collected.

The severity of the depressive episode was retrospectively determined based upon medical record review of the clinical documentation using the Clinical Global Impression Severity scale (CGI-S) [24] at the time of SSRI initiation and at follow-up (first psychiatric note following 12 weeks of treatment). The CGI-S is a clinician-rated scale with scores ranging from 1 to 7 (1 = not at all ill; 2 = borderline mentally ill; 3 = mildly ill; 4 = moderately ill; 5 = markedly ill; 6 = severely ill; 7 = extremely ill).

Treatment response was coded on the Clinical Global Impression Improvement scale (CGI-I) anchored to change in depression symptoms over the initial 12 weeks of treatment. The CGI was developed for use in National Institutes of Mental Health-sponsored clinical trials to provide a brief, stand-alone assessment of the clinician's view of the patient's global functioning before and after starting a study medication. The CGI has been shown to correlate well with standard research medication efficacy scales across a wide range of psychiatric conditions and the CGI is used in virtually all FDA-regulated psychiatric trials [25]. The CGI-I is a clinician-rated scale with scores ranging from 1 to 7 (1 = very much improved; 2 = much improved; 3 = minimally improved; 4 = no change; 5 = minimally worse; 6 = much worse; 7 = very much worse). The CGI-S and CGI-I ratings were assigned by a board-certified psychiatrist with expertise in treating adults with developmental disabilities who was not the treating psychiatrist (RPT).

2.3. Statistical Analysis

Categorical variables were summarized using frequencies and percentages. Continuous variables were summarized using means, standard deviations (SDs), medians, and ranges. Treatment response was defined using the CGI-I score, with values ≤ 2 (very much improved or much improved) corresponding to response and values ≥ 3 (minimally improved or worse) corresponding to non-response. Patients were classified as having attained remission and recovery if at least three weeks or four months, respectively, of minimal depressive symptoms [26] had been attained at the time of the most recent psychiatry follow-up visit. Ninety-five percent confidence intervals (CIs) for percentages were calculated using Wilson's method. Time to discontinuation of SSRI medication was characterized using a Kaplan-Meier survival curve plotted using Stata software (version 14). Other data analysis was conducted using SAS software (version 9.4).

3. Results

3.1. Study Flow

Records of 46 patients were identified for review, of which 11 were included in the study. Figure 1 shows the number of patients successively meeting each of the eligibility criteria. Complete data were collected for all 11 patients included in the study.

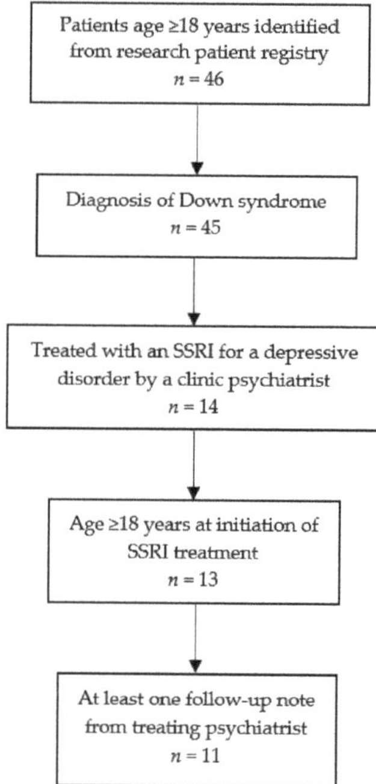

Figure 1. Flow diagram of identification of patients in the study. SSRI: selective serotonin reuptake inhibitor.

3.2. Demographic and Clinical Characteristics of the Sample

Demographic and clinical characteristics of the sample at the time of initiation of SSRI treatment are presented in Table 1. Seven of the 11 patients (64%) were male, and 10 (91%) were White. Age ranged from 18 to 46 years. Three (27%) had a comorbid psychiatric diagnosis in addition to the depressive disorder, and all had one or more comorbid medical diagnoses. Five (45%) had a history of treatment with any psychiatric medication, and three (27%) had a history of treatment with an SSRI.

Table 1. Demographic and Clinical Characteristics of Patients at the Time of SSRI Initiation.

	Eligible for Inclusion *n* = 11
Sex, *n* (%)	
Female	4 (36%)
Male	7 (64%)
Age, mean (SD; range in years)	27.2 (7.7; 18–46)
Race [1], *n* (%)	
White	10 (91%)
Other	1 (9%)
Intellectual Disability, *n* (%)	
Any intellectual disability	11 (100%)
Mild	2 (18%)
Moderate	8 (73%)
Severe	1 (9%)
Language Ability	
Full sentences	7 (64%)
Phrase speech	3 (27%)
Single words	1 (9%)
Age at onset of first depressive episode, mean (SD; range in years)	21.6 (10.1; 5–44)
Comorbid psychiatric diagnoses, *n* (%)	
Any diagnosis	3 (27%)
Generalized anxiety disorder	2 (18%)
Alzheimer's disease	1 (9%)
Comorbid medical diagnoses [2], *n* (%)	
Any diagnosis	11 (100%)
Congenital heart disease	7 (64%)
Obstructive sleep apnea	4 (36%)
Hypothyroidism	3 (27%)
Prior psychiatric medications, *n* (%)	
Any prior psychiatric medication	5 (45%)
SSRI [3]	3 (27%)
Second-generation antipsychotic	2 (18%)
Buspirone	2 (18%)
Donepezil	1 (9%)

[1] As categorized by the research patient registry; categories include Asian, Black or African American, Hispanic, White, and Other. [2] Less frequent comorbid medical diagnoses were constipation (*n* = 2), eczema (*n* = 2), GERD (*n* = 2), hearing loss (*n* = 2), hyperopia (*n* = 2), cataract (*n* = 1), celiac disease (*n* = 1), esotropia (*n* = 1), head trauma (*n* = 1), onychomycosis (*n* = 1), otitis media (*n* = 1), psoriasis (*n* = 1), and type 1 diabetes (*n* = 1). [3] SSRI: selective serotonin reuptake inhibitor; GERD: gastroesophageal reflux disease.

3.3. Characteristics of Depressive Episodes and SSRI Treatment

Selective serotonin reuptake inhibitor treatment was initiated between 2013 and 2020. Characteristics of the depressive episodes and treatment are presented in Table 2. Depression severity at the time of SSRI initiation as rated by the CGI-S ranged from mildly ill (CGI-S = 3) to severely ill (CGI-S = 6), with most patients rated as moderately ill (CGI-S = 4). Fluoxetine was the most commonly prescribed SSRI (*n* = 8). Two patients (18%) were prescribed sertraline, and one patient (9%) was prescribed escitalopram. Five patients (45%) were taking one or more concomitant psychiatric medications, and 10 (91%) were receiving one or more non-medication treatments (nine patients participated in a day program, and five patients received individual psychotherapy). Two patients (18%) had depression with psychotic features.

Table 2. Characteristics of Depressive Episodes and Treatment.

	Eligible for Inclusion $n = 11$
Depressive Episode	
CGI-S score, *n* (%)	
3: Mildly ill	1 (9%)
4: Moderately ill	6 (55%)
5: Markedly ill	2 (18%)
6: Severely ill	2 (18%)
Suicidal ideation at baseline, *n* (%)	1 (9%)
First depressive episode, *n* (%)	7 (64%)
Psychotic features, *n* (%)	2 (18%)
Treatment	
SSRI, *n* (%)	
Fluoxetine	8 (73%)
Initial dose in mg, mean (SD; range)	4.9 (0.4; 4.0–5.0)
Maximal dose in mg, mean (SD; range)	25.5 (18.7; 14.0–70.0)
Sertraline [2]	2 (18%)
Escitalopram [2]	1 (9%)
Concomitant medication treatments [1], *n* (%)	
Any medication treatment	5 (45%)
Second-generation antipsychotic	3 (27%)
Buspirone	2 (18%)
Concomitant non-medication treatments, *n* (%)	
Any non-medication treatment	10 (91%)
Day program	9 (82%)
Individual psychotherapy	5 (45%)

[1] One patient each was taking an alpha-2 agonist, N-acetylcysteine, rivastigmine, and trazodone. [2] See Table 3 for dosage information. CGI-S: Clinical Global Impression Severity scale. SSRI: selective serotonin reuptake inhibitor.

Table 3. Summary of 12-week and Long-Term Treatment Course.

Patient	SSRI	CGI-S (Baseline)	CGI-I (12 Weeks)	Follow-up Duration	Long-Term Treatment Course
1	sertraline	5	5	37 weeks	Depression became more severe and psychotic features emerged over first 12 weeks of treatment. Ultimately did well on sertraline 62.5 mg per day and aripiprazole 6 mg per day.
2	sertraline	4	2	6.7 years	Depression continued to improve after 12 weeks with gradual upward dose titration. Doing well on sertraline 125 mg per day at last follow-up.
3	escitalopram	3	1	2.1 years	Depression responded well to escitalopram 12.5 mg per day for ~2 years, but anxiety persisted. Higher doses of escitalopram were associated with fatigue, tearfulness, and irritability. Escitalopram was switched to another SSRI, which also caused tearfulness. The patient was doing well without medications for several months at last follow-up.
4	fluoxetine	5	1	2.0 years	Continued to do well on fluoxetine 14 mg per day at last follow up.
5	fluoxetine	4	2	4.5 years	Continued to do well on fluoxetine 15 mg per day and aripiprazole 2 mg per day at last follow-up. Did not tolerate aripiprazole taper due to irritability.

Table 3. *Cont.*

Patient	SSRI	CGI-S (Baseline)	CGI-I (12 Weeks)	Follow-up Duration	Long-Term Treatment Course
6	fluoxetine	6	3	2.6 years	Experienced 25–30% improvement on fluoxetine 55 mg per day. Adjunctive buspirone 7.5 mg BID resulted in remission after six months of treatment.
7	fluoxetine	4	2	4.5 years	After six months of fluoxetine 20 mg per day, experienced mania, and fluoxetine was discontinued. Ultimately did well on carbamazepine 300 mg BID and guanfacine 0.5 mg BID.
8	fluoxetine	4	2	24 weeks	Experienced 50% improvement on fluoxetine 20 mg per day. When fluoxetine was increased to 25 mg, experienced agitation, irritability, and anger. Follow-up data after fluoxetine was tapered is not available.
9	fluoxetine	6	2	33 weeks	Experienced moderate improvement on fluoxetine 15 mg per day.
10	fluoxetine	4	2	4.5 years	Depression responded to fluoxetine 20 mg per day. After two years of stability, fluoxetine was tapered and depression recurred. Symptoms resolved when fluoxetine was increased back to 20 mg per day.
11	fluoxetine	4	1	3.1 years	After three years of stability on fluoxetine 30 mg, fluoxetine was tapered and discontinued without return of depression.

SSRI: selective serotonin reuptake inhibitor; CGI-S: Clinical Global Impression Severity scale; CGI-I: Clinical Global Impression Improvement scale.

3.4. Response to SSRI Treatment

All 11 patients completed at least a 12-week course of SSRI treatment. The median duration of the follow-up period from time of SSRI treatment initiation to the end of the 12-week treatment course was 14.4 weeks, with a range of 12.0–33.0 weeks. Clinical Global Impression ratings before and after 12 weeks of treatment are presented in Figure 2. Based on the CGI-I rating for the soonest psychiatric visit after 12 weeks of treatment, nine of the 11 patients responded (response rate 82%; 95% CI 52%, 95%), with three patients rated as very much improved (CGI-I = 1) and six patients rated as much improved (CGI-I = 2). Eight of the 11 patients (73%) were rated as mildly ill or less at follow-up (CGI-S $\leq$ 3), with five patients rated as mildly ill (CGI-S = 3), two patients rated as borderline mentally ill (CGI-S = 2), and one patient rated as not at all ill (CGI-S = 1). One of the remaining three patients, taking fluoxetine, was much improved (CGI-I = 2) but still moderately ill (CGI-S = 5); one taking fluoxetine was minimally improved (CGI-I = 3) and still moderately ill (CGI-S = 4); and one taking sertraline was minimally worse (CGI-I = 5) and severely ill (CGI-S = 6). Of the three patients who had previously been prescribed an SSRI, two of the patients were responders (CGI-I = 1 or 2).

Table 3 summarizes the long-term treatment course beyond the initial 12 weeks of treatment with an SSRI. The median duration of time after initiation of the SSRI covered by record review was 2.1 years, with a range of 24 weeks to 6.7 years. Seven patients (64%) had a sustained positive response to the SSRI until the most recent follow-up, and one patient (9%) had a positive response to fluoxetine and was able to discontinue it after three years of stability without a return of depression. Three patients (27%) had an initially positive response to the SSRI but later experienced either behavioral activation or mania with sustained treatment. Of these three patients, two patients (18%) experienced behavioral activation (irritability, anger, and/or agitation) when the SSRI was titrated to a certain dose (escitalopram 15 mg daily and fluoxetine 20 mg daily). The third patient experienced a manic episode after six months of treatment with fluoxetine 15–20 mg daily. Three patients were optimally managed on the SSRI plus an adjunctive medication, either aripiprazole (a second-generation antipsychotic) [$n = 2$], or buspirone (a serotonin-$_{1A}$ partial agonist) [$n = 1$].

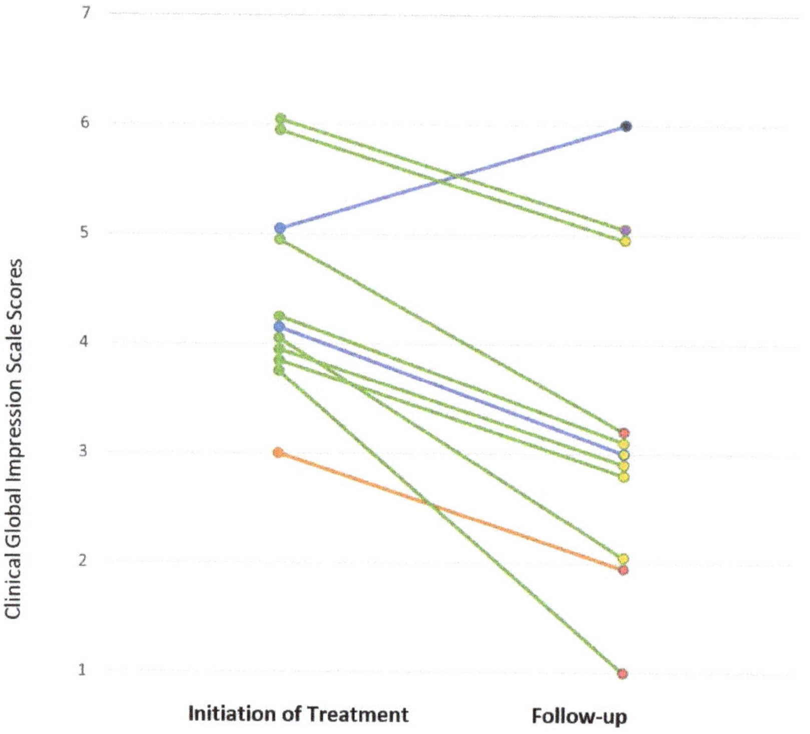

Figure 2. Clinical Global Impression Severity Scale (CGI-S) scores at the initiation of SSRI treatment and at first follow-up visit following 12 weeks of treatment. Each line represents the change in CGI-S score for one patient. Line color corresponds to SSRI: green = fluoxetine, blue = sertraline, orange = escitralopram. Color of marker at follow-up corresponds to CGI-Improvement scale (CGI-I) score: pink = very much improved (CGI-I = 1), yellow = much improved (CGI-I = 2), purple = minimally improved (CGI-I = 3), gray=minimally worse (CGI-I = 5).

Figure 3 presents results on the duration of medication use. Four patients (36%) discontinued use of SSRI medications in the period covered by psychiatric notes in the medical record: three (at 24 weeks, 31 weeks, and 2.1 years, respectively) because of both loss of effectiveness and difficulty tolerating the medication and one (at 3.1 years) because the medication was no longer needed. The duration of medication use at the time of the most recent psychiatric note for the seven patients who remained on medication ranged from 33 weeks to 6.7 years. For the two patients remaining on sertraline, the doses at the time of last follow-up note were 62.5 mg and 125 mg daily. For the five patients remaining on fluoxetine, mean (SD) final dose was 26.8 (24.3) mg per day, with a range of 14.0–70.0 mg per day. Overall, 9/11 patients (82%; 95% CI 52%, 95%) achieved remission ($\geq$3 weeks of minimal depressive symptoms), and 8/11 patients (73%; 95% CI 43%, 90%) achieved recovery ($\geq$4 months of minimal depressive symptoms) based upon the most recent psychiatric follow-up note.

3.5. Adverse Effects

Four of the 11 patients (36%; 95% CI 15%, 65%) had adverse effects reported in their psychiatric notes. Daytime sedation and anger were reported for two patients (18%) each, and weight gain, behavioral activation, irritability, anxiety, and mania were reported for one patient (9%) each. There was no indication of increased suicidal thinking, intent, plan, or attempts in any of the psychiatric follow-up notes.

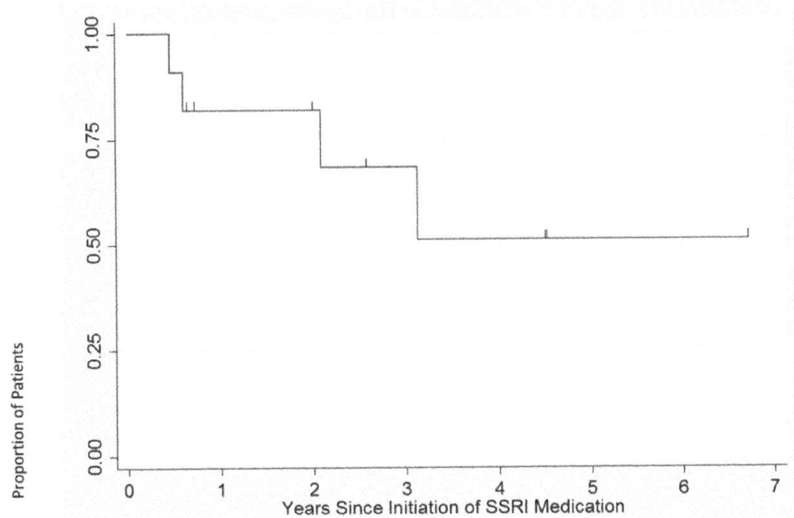

Figure 3. Proportion of patients remaining on SSRI medication over time. Tick-marks correspond to time of last psychiatric note for patients who remained on an SSRI at most recent psychiatric visit.

4. Discussion

This naturalistic, retrospective study evaluated the real-world effectiveness and tolerability of SSRIs for the treatment of depressive disorders in a small sample of adults with DS. This study reports on the largest sample of individuals with DS treated with SSRIs to date. It also offers insight into how SSRIs are used in a long-term naturalistic tertiary care clinical setting. This study informs on several issues that have not been previously reported in adults with DS, including the clinical characteristics of those receiving treatment with SSRIs, the initial 12-week response rate to SSRIs, long-term tolerability of SSRIs, long-term response, and adverse effects. Selective serotonin reuptake inhibitors are the most commonly prescribed first-line medications for depression in the general population. Previous studies of pharmacotherapy for depression in adults with DS have predominantly reported on TCAs, which are now reserved for treatment-refractory depression in modern psychopharmacology due to the potential for life-threatening adverse effects. The results from this study suggest that the treatment of depression in DS with SSRIs was overall well tolerated and safe, and was associated with clinically significant improvement in symptoms of depression in most adults with DS; however, prospective randomized controlled trials are needed to provide conclusive evidence.

In this study, a majority (82%; 95% CI 52%, 95%) of patients responded to SSRI therapy (CGI-I $\leq$ 2). This response rate is similar to the data reported in a retrospective study conducted by Myers et al. [18]. Myers et al. reviewed the records of 164 adults treated as outpatients from 1979–1989 in the Down Syndrome Program at the Child Development Center at Rhode Island Hospital and identified nine adults with a depressive disorder, three of whom were treated with an SSRI (fluoxetine). All three patients (100%) who received treatment with fluoxetine (mean dosage: 47 mg per day) were deemed responders [18]. No other reports on the use of SSRIs for depression in DS have been published. Data from our study also indicate that the majority of patients achieved long term response, as indicated by the 82% (95% CI 52%, 95%) remission or recovery rate. This high rate of remission or recovery demonstrates the overall treatability of depression in DS, underscoring the importance of accurate case detection and availability of treatment.

Findings from this study also suggest that psychotic features may be commonly associated with depression in DS. Two of the 11 patients (18%; 95% CI 5%, 48%) in this sample experienced depression with psychotic features. One patient complained of hearing voices, was observed to be responding to internal stimuli, and exhibited disorganized behaviors, such as trying to leave the house in the middle of the night without clothing. The other patient experienced auditory hallucinations of derogatory content, which resulted in distress manifesting as pushing, screaming, and looking for knives as well as paranoid delusions of people breaking into the house. Both patients were successfully treated with a combination of fluoxetine and aripiprazole (a second-generation antipsychotic). Previous literature also supports an elevated rate of depression with psychotic features in DS. A study including 49 adolescents and young adults (13–29 years) treated in specialized psychiatric clinics reported that 8% of patients with DS had a history of depression with psychotic features [11]. Another study comparing the prevalence of obstructive sleep apnea in adults with DS with or without depression reported that 9/28 (32%) patients in the depression group had accompanying psychotic features [27].

Patients in this study generally responded to lower dosages of SSRIs than is typically used in the general population [28]. The mean starting dosage of fluoxetine was 4.9 mg daily and the mean maximal dosage was 25.5 mg daily, which approximates the recommended starting dosage of 20 mg daily in the general adult population [28]. This dosing strategy differs from the three cases reported by Myers et al., in which more typical adult dosing was used (optimal dosages of 20 mg, 40 mg, and 80 mg daily) [18]. Especially in light of the observation that two patients (18%) in our study experienced behavioral activation at a certain dosage threshold (fluoxetine 20 mg and escitalopram 15 mg daily) and a third experienced a manic episode after six months of treatment with fluoxetine 15–20 mg daily, the results from this study support a more conservative dosing strategy.

Although SSRIs were generally well tolerated, four of the 11 patients (36%; 95% CI 15%, 65%) had adverse effects reported in their psychiatric notes. The most common adverse effects were daytime sedation and anger which were reported in two patients each. The majority of patients followed tolerated long-term use, with three patients (27%) discontinuing the medication due to both loss of effectiveness and difficulty tolerating the medication related to either behavioral activation ($n = 2$) or mania ($n = 1$). It is of interest that no patients in this sample reported gastrointestinal side effects, which is the most common reason for early discontinuation of SSRIs [20]. The use of lower dosages of SSRIs and a slower titration schedule are known to mitigate SSRI-related side effects and the dosage titration pattern observed in this study may explain why this was observed.

Limitations

The results are subject to several limitations, including the small sample size, chart review nature of the analysis, which lacks a placebo or control group and standardized rating scales administered at the time of treatment and potential confounding factors associated with the concomitant use of other psychiatric medications and nonmedication treatments. The primary outcome measure of this study was the CGI, anchored to depression. A limitation of using a global rating for treatment response is the inability to determine which symptoms of depression were responsive to SSRIs. Because no standard depression severity rating scales have been used in patients with DS, an overall clinical impression rating scale rated by a single rater with expertise in treating adults with neurodevelopmental disabilities provides preliminary, yet clinically relevant, information to the literature, which can be used as a basis for developing prospective studies to include and assess depression rating scales in this population. The prevalence of side effects reported in this study may be underestimated, as they were not collected in a systematic fashion and patients with DS may be less able to accurately report side effects due to the cognitive and communication limitations associated with the syndrome. Additionally, although the duration of the initial treatment period was pre-determined to be 12 weeks, the median time of clinical follow-up was 14.4 weeks, with a range of 12–33 weeks. While the sample size was limited to 11 patients, this report includes the largest sample of patients with DS treated with SSRIs for depression. In addition, the diagnosis of depressive disorders was based on expert clinical evaluation rather than through the use of standardized assessment tools. The sample may be biased toward individuals with more serious psychopathology, given that it was completed at a tertiary care center and many individuals with DS and milder depression may not be identified and/or referred to psychiatry. Many of these limitations are a function of the retrospective, naturalistic aspects of the study. This study design does offer the advantage of providing insight into the effectiveness, tolerability, and safety of SSRIs in real-world clinical practice.

5. Conclusions

Overall, most adults with DS and depression responded to a 12-week course of SSRI treatment, and some tolerated long-term use. Although seven of the 11 patients in the study had no adverse effects reported in their psychiatric notes, several experienced behavioral activation and one experienced mania. We believe our findings warrant a future prospective randomized placebo-controlled study of SSRIs in adults with DS.

Author Contributions: Conceptualization, R.P.T., M.L.P., C.J.M. and C.T.R.; methodology, R.P.T., M.L.P., C.J.M. and C.T.R.; formal analysis, C.T.R.; investigation, R.P.T., M.L.P., C.J.M., C.T. and C.T.R.; data curation, R.P.T., C.T.; writing—original draft preparation, R.P.T.; writing—review and editing, R.P.T., M.L.P., C.J.M., C.T. and C.T.R. All authors have read and agreed to the published version of the manuscript.

Funding: This work was funded, in part, by the Nancy Lurie Marks Family Foundation.

Institutional Review Board Statement: The study was conducted according to the guidelines of the Declaration of Helsinki, and approved by the Institutional Review Board of Mass General Brigham (protocol code 2021P000640 and March 12, 2021).

Informed Consent Statement: Patient consent was waived due because requiring consent would lead to years of delay, incomplete information, biased analyses, and would result in less utility of the data in that it will not be possible to perform effective hypotheses testing, quality improvement analyses and population management without access to the full data set. No experimental intervention was involved. Patients underwent clinical assessments and receive standard of care as determined by the patients' physician.

Data Availability Statement: The data presented in this study are available in this article.

Conflicts of Interest: The authors declare no conflict of interest.

Disclosures: Precidiag—Consultant (C.J.M.); Receptor Life—Consultant (C.J.M.); Oxford University Press—Royalties (C.J.M.); Springer Publishing—Royalties/Editorial Role (C.J.M.).

References

1. CDC. Facts about Down Syndrome. Available online: https://www.cdc.gov/ncbddd/birthdefects/downsyndrome.html (accessed on 25 June 2020).
2. Mai, C.T.; Isenburg, J.L.; Canfield, M.A.; Meyer, R.E.; Correa, A.; Alverson, C.J.; Lupo, P.J.; Riehle-Colarusso, T.; Cho, S.J.; Aggarwal, D.; et al. National population-based estimates for major birth defects, 2010–2014. *Birth Defects Res.* **2019**, *111*, 1420–1435. [CrossRef]
3. Ivan, D.L.; Cromwell, P. Clinical practice guidelines for management of children with Down syndrome: Part, I. *J. Pediatric Health Care* **2014**, *28*, 105–110. [CrossRef]
4. Gath, A.; Gumley, D. Behavior problems in retarded children with special reference to Down's syndrome. *Br. J. Psychiatry* **1986**, *149*, 156–161. [CrossRef] [PubMed]
5. McCarthy, J.; Boyd, J. Psychopathology and young people with down's syndrome: Childhood predictors and adult outcome of disorder. *J. Intellect. Disabil. Res.* **2001**, *45*, 99–105. [CrossRef] [PubMed]
6. Palumbo, M.L.; McDougle, C.J. Pharmacotherapy of down syndrome. *Expert Opin. Pharmacother.* **2018**, *19*, 1875–1889. [CrossRef] [PubMed]
7. Visootsak, J.; Sherman, S. Neuropsychiatric and behavioral aspects of trisomy 21. *Curr. Psychiatry Rep.* **2007**, *9*, 135–140. [CrossRef]
8. Dykens, E.M. Psychiatric and behavioral disorders in persons with Down syndrome. *Ment. Retard. Dev. Disabil. Res. Rev.* **2007**, *13*, 272–278. [CrossRef] [PubMed]
9. Myers, B.A.; Pueschel, S.M. Psychiatric disorders in persons with down syndrome. *J. Nerv. Ment. Dis.* **1991**, *179*, 609–613. [CrossRef] [PubMed]

10. Cooper, S.A.; Collacott, R.A. Clinical features and diagnostic criteria of depression in Down's syndrome. *Br. J. Psychiatry* **1994**, *165*, 399–403. [CrossRef]

11. Dykens, E.M.; Shah, B.; Davis, B.; Baker, C.; Fife, T.; Fitzpatrick, J. Psychiatric disorders in adolescents and young adults with Down syndrome and other intellectual disabilities. *J. Neurodev. Disord.* **2015**, *7*, 9. [CrossRef]

12. Mantry, D.; Cooper, S.A.; Smiley, E.; Morrison, J.; Allan, L.; Williamson, A.; Finlayson, J.; Jackson, A. The prevalence and incidence of mental ill-health in adults with Down syndrome. *J. Intellect. Disabil. Res.* **2008**, *52*, 141–155. [CrossRef]

13. Startin, C.M.; D'Souza, H.; Ball, G.; Hamburg, S.; Hithersay, R.; Hughes, K.M.O.; Massand, E.; Karmiloff-Smith, A.; Thomas, M.S.C.; Strydom, A.; et al. Health comorbidities and cognitive abilities across the lifespan in down syndrome. *J. Neurodev. Disord.* **2020**, *12*, 4. [CrossRef] [PubMed]

14. Zammit, S.; Allebeck, P.; David, A.S.; Dalman, C.; Hemmingsson, T.; Lundberg, I.; Lewis, G. A longitudinal study of premorbid IQ score and risk of developing schizophrenia, bipolar disorder, severe depression, and other nonaffective psychoses. *Arch. Gen. Psychiatry* **2004**, *61*, 354–360. [CrossRef]

15. Seidl, R.; Kaehler, S.T.; Prast, H.; Singewald, N.; Cairns, N.; Gratzer, M.; Lubec, G. Serotonin (5-HT) in brains of adult patients with Down syndrome. *J. Neural Transm. Suppl.* **1999**, *57*, 221–232. [CrossRef]

16. Walker, J.C.; Dosen, A.; Buitelaar, J.K.; Janzing, J.G.E. Depression in Down syndrome: A review of the literature. *Res. Dev. Disabil.* **2011**, *32*, 1432–1440. [CrossRef] [PubMed]

17. Storm, W. Differential diagnosis and treatment of depressive features in Down's syndrome: A case illustration. *Res. Dev. Disabil.* **1990**, *11*, 131–137. [CrossRef]

18. Myers, B.A.; Pueschel, S.M. Major depression in a small group of adults with Down syndrome. *Res. Dev. Disabil.* **1995**, *16*, 285–299. [CrossRef]

19. Szymanski, L.; Biederman, J. Depression and anorexia nervosa of persons with Down syndrome. *Am. J. Ment. Defic.* **1984**, *89*, 246–251.

20. Schatzberg, A.; DeBattista, C. *Manual of Clinical Psychopharmacology*, 8th ed.; American Psychiatric Publishing, Inc.: Arlington, VA, USA, 2015.

21. Kennedy, S.H.; Lam, R.W.; McIntyre, R.S.; Tourjman, S.V.; Bhat, V.; Blier, P.; Hasnain, M.; Jollant, F.; Levitt, A.J.; MacQueen, G.M.; et al. Canadian Network for Mood and Anxiety Treatments (CANMAT) 2016 clinical guidelines for the management of adults with major depressive disorder: Section 3. Pharmacological treatments. *Can. J. Psychiatry* **2016**, *61*, 540–560. [CrossRef] [PubMed]

22. Denee, T.; Kerr, C.; Ming, T.; Wood, R.; Tritton, T.; Middleton-Dalby, C.; Massey, O.; Desai, M. Current treatments used in clinical practice for major depressive disorder and treatment resistant depression in England: A retrospective database study. *J. Psychiatr. Res.* **2021**, *139*, 172–178. [CrossRef] [PubMed]

23. Cascade, E.; Kalali, A.H.; Kennedy, S.H. Real-world data on SSRI antidepressant side effects. *Psychiatry* **2009**, *6*, 16–18.

24. CGI—Clinical Global Impressions Scale. Available online: https://eprovide.mapi-trust.org/instruments/clinical-global-impressions-scale (accessed on 23 June 2020).

25. Busner, J.; Targum, S.D. The clinical global impressions scale: Applying a research tool in clinical practice. *Psychiatry* **2007**, *4*, 28. [PubMed]

26. Rush, A.J.; Kraemer, H.C.; Sackeim, H.A.; Fava, M.; Trivedi, M.H.; Frank, E.; Ninan, P.T.; Thase, M.E.; Gelenberg, A.J.; Kupfer, D.J.; et al. Report by the ACNP Task Force on response and remission in major depressive disorder. *Neuropsychopharmacology* **2006**, *31*, 1841–1853. [CrossRef] [PubMed]

27. Capone, G.T.; Aidikoff, J.M.; Taylor, K.; Rykiel, N. Adolescents and young adults with down syndrome presenting to a medical clinic with depression: Co-morbid obstructive sleep apnea. *Am. J. Med. Genet. Part A* **2013**, *161*, 2188–2196. [CrossRef] [PubMed]

28. Stahl, S. *Essential Psychopharmacology: The Prescriber's Guide*; Cambridge University Press: New York, NY, USA, 2005.

Article

Cognitive and Behavioral Domains That Reliably Differentiate Normal Aging and Dementia in Down Syndrome

Jordan P. Harp [1,*], Lisa M. Koehl [1], Kathryn L. Van Pelt [2], Christy L. Hom [3], Eric Doran [4], Elizabeth Head [5], Ira T. Lott [6] and Frederick A. Schmitt [7]

[1] Department of Neurology, University of Kentucky, Lexington, KY 40506, USA; lisa.mason@uky.edu
[2] Sanders-Brown Center on Aging, University of Kentucky, Lexington, KY 40506, USA; vanpelt.kathryn@gmail.com
[3] Department of Psychiatry and Human Behavior, University of California, Irvine, CA 92697, USA; homc@uci.edu
[4] Department of Pediatrics, School of Medicine, University of California, Irvine, CA 92697, USA; edoran@hs.uci.edu
[5] Department of Pathology & Laboratory Medicine, University of California, Irvine, CA 92697, USA; heade@uci.edu
[6] Departments of Neurology and Pediatrics, University of California, Irvine, CA 92697, USA; itlott@hs.uci.edu
[7] Department of Neurology and Sanders-Brown Center on Aging, University of Kentucky, Lexington, KY 40506, USA; fascom@uky.edu
* Correspondence: jordanharp@uky.edu

Abstract: Primary care integration of Down syndrome (DS)-specific dementia screening is strongly advised. The current study employed principal components analysis (PCA) and classification and regression tree (CART) analyses to identify an abbreviated battery for dementia classification. Scale- and subscale-level scores from 141 participants (no dementia $n = 68$; probable Alzheimer's disease $n = 73$), for the Severe Impairment Battery (SIB), Dementia Scale for People with Learning Disabilities (DLD), and Vineland Adaptive Behavior Scales—Second Edition (Vineland-II) were analyzed. Two principle components (PC1, PC2) were identified with the odds of a probable dementia diagnosis increasing 2.54 times per PC1 unit increase and by 3.73 times per PC2 unit increase. CART analysis identified that the DLD sum of cognitive scores (SCS < 35 raw) and Vineland-II community subdomain (<36 raw) scores best classified dementia. No significant difference in the PCA versus CART area under the curve (AUC) was noted (D(65.196) = −0.57683; $p = 0.57$; PCA AUC = 0.87; CART AUC = 0.91). The PCA sensitivity was 80% and specificity was 70%; CART was 100% and specificity was 81%. These results support an abbreviated dementia screening battery to identify at-risk individuals with DS in primary care settings to guide specialized diagnostic referral.

Keywords: Down syndrome; dementia; cognition; functional independence; neuropsychological assessment; primary care; screening

Citation: Harp, J.P.; Koehl, L.M.; Pelt, K.L.V.; Hom, C.L.; Doran, E.; Head, E.; Lott, I.T.; Schmitt, F.A. Cognitive and Behavioral Domains That Reliably Differentiate Normal Aging and Dementia in Down Syndrome. *Brain Sci.* **2021**, *11*, 1128. https://doi.org/10.3390/brainsci11091128

Academic Editor: Corrado Romano

Received: 15 July 2021
Accepted: 22 August 2021
Published: 25 August 2021

1. Introduction

Down syndrome (DS), a genetic condition caused predominantly by the triplication of chromosome 21, is highly associated with the development of Alzheimer's disease (AD) [1]. Chromosome 21 includes the amyloid precursor protein (APP) gene, and triplication results in overexpression of APP and related proteins, accelerating the accumulation of misfolded amyloid in the brain [2–4]. Additional AD risk factors are also associated with DS including a higher propensity for neuroinflammation, oxidative damage, sleep apnea, and reduced cognitive reserve due to premorbid intellectual disability [1,5–7]. Indeed, AD pathological changes have been documented in adults with DS as young as 20 years, and nearly all adults with DS show the amyloid plaques and neurofibrillary tangles associated with AD by 40 years of age [8–10].

DS is associated with different physical morphology, intellectual disabilities, and reduced lifespan compared to the typically developing population. Associated health problems include atlantoaxial instability, musculoskeletal and dental conditions, congenital heart disease, hematologic conditions, obesity, hypothyroidism, obstructive sleep apnea, impaired hearing and vision, and overall increased functional dependence due to behavioral, psychiatric, and intellectual impairments [7,11–14]. Advances in medical management of these co-morbidities have lowered mortality from early-life conditions, but one consequence of lengthened lifespan is that more individuals with DS now survive to the age of risk for AD [15].

Due to the need for preventative care and ongoing management of chronic health conditions associated with DS, health professionals and advocacy groups recommend the integration of DS-specific care in primary care settings [14,16,17]. Healthcare systems have made progress toward this end, but there is a need for improvement [11,18]. Cognitive screening and monitoring for dementia is particularly difficult, as cognitive measurement is complicated by pre-existing intellectual disability (ID), large inter- and intra-individual variability in cognition and behavior, tolerability of testing methods, and the lack of an identified "gold standard" neurocognitive battery, even for research purposes [19,20]. Moreover, neurocognitive tests are not feasible in primary care settings due to the lengthy procedures and specialized training needed for the interpretation of comprehensive evaluations.

In recent studies, our group has sought to establish an evidence base for abbreviated neurobehavioral examination procedures appropriate for in-office dementia monitoring by community practitioners caring for patients with DS [21]. Performance measures in our long-term cohort studies include the Brief Praxis Test (BPT) [22] and the Severe Impairment Battery (SIB) [23]. Informant measures included the Dementia Questionnaire for People with Learning Disabilities (DLD) [24] and Vineland Adaptive Behavior Scales-Second Edition (Vineland-II) [25]. The BPT, SIB, and DLD have all been used in early DS clinical trials assessing the effects of anticholinesterase therapy [26] as well as antioxidants [27,28]. Moreover, the SIB has long been validated as a cognitive measure for severe impaired individuals with AD [29]. The Vineland-II has been widely used and validated in the DS population [30–32] and adaptive behavior decline is a diagnostic criterion for AD, necessitating the inclusion of this type of measure in this study. These measures were selected at the outset of the two parent cohorts from which the present data are drawn, and target the domains of cognition (SIB, DLD), praxis (BPT), and functional independence (DLD, Vineland-II) that underlie both NINCDS-ADRDA and DSM-IV criteria for dementia/major neurocognitive disorder.

The present study seeks to further identify the key components that are useful for dementia detection through three aims:

- Aim 1: to identify the underlying components of a cognitive battery that was used to assess functioning in domains commonly affected by AD.
- Aim 2: to select the minimum necessary individual items or subscales using CART analysis to create an abbreviated battery for classifying AD status.
- Aim 3: to compare the classification accuracy between the two methods: components from the full battery vs. the abbreviated battery.

2. Materials and Methods

2.1. Description of Sample

The current study combines participants from cohorts at two different sites: the University of Kentucky and the University of California, Irvine (UCI). The University of Kentucky Aging and Down Syndrome (ADS) study is a longitudinal cohort of aging individuals with DS. For the purpose of the current study, only the baseline visit was used for 88 participants. Twenty-nine of the original one hundred and seventeen participants in the overall ADS study were unable to contribute data for the present analysis, predominantly due to inability to engage in testing because of advanced dementia. The University of Kentucky ADS cohort recruited individuals with DS between 25 and 64 years of age. From the UCI cohort, only the baseline visit was used for 53 participants. One of the original 54 participants in the overall UCI study was unable to contribute data for the present analysis due to inability to engage in testing because of advanced dementia. The UCI cohort included people between 43 and 58 years of age.

A karyotype diagnosis of trisomy 21 (full or mosaic) or the Robertsonian translocation form of DS was required. Baseline levels of ID were determined by caregiver report of prior evaluation results or by a review of records when available. Other requirements for study inclusion included a stable medical condition for at least 3 months prior to the study and to have an absence of systemic disorders that might confound a diagnosis of dementia. Medication usage including psychotropic and Parkinsonian drugs was required to be stable for 3 months prior to study, and English-speaking skills were required to facilitate neuropsychological testing.

Research procedures were independently reviewed and approved by the University of Kentucky Institutional Review Board and the UCI Institutional Review Board. Participants completed approved protocols for informed consent or assent with guardian or legally authorized representative approval.

2.2. Description of Measures

Both sites administered a combination of performance and informant measures that have been used with adults with DS. Performance measures included the BPT and SIB, and informant measures included the DLD and Vineland-II.

The BPT is a 20-item measure of dyspraxia that minimizes verbal demands in favor of simple behavioral output. Low scores on the BPT indicate severe dyspraxia.

The SIB utilizes one-step commands and gestural cues, and allows for non-verbal responses and partially correct responses in order to assess cognition in individuals with severe dementia. The SIB yields a total score along with six major subscales for attention, orientation, language, memory, visuospatial ability, and construction, with additional scores for orientation to name, praxis, and social interaction. Lower scores indicate more severe deficits.

The DLD is a 50-item informant questionnaire measuring behavioral and cognitive dysfunction. The DLD yields three scores: (1) sum of cognitive score (SCS), measuring short-term memory, long-term memory, and spatial/temporal orientation; (2) sum of social score (SOS), measuring speech, practical skills, mood, activity/interest, and behavioral disturbance; and (3) a total score that combines the SCS and SOS. DLD raters for the current study were caregivers and/or legal guardians responsible for the daily care of the participants either at home or an assisted living facility. Higher scores on the DLD indicate more severe impairment.

The Vineland-II is an informant-based measure covering domains of communication, daily living skills, socialization, motor skills, and maladaptive behavior. The Vineland-II provides a composite score reflecting an individual's overall adaptive behavior functioning, called the Adaptive Behavior Composite (ABC). The Vineland-II is administered by a trained interviewer to the parent or caregiver.

2.2.1. Consensus Diagnosis

AD diagnosis, based on NINCDS-ADRDA or DSM-IV criteria [33,34], was made at each site using a consensus process involving a neurologist and a psychologist. The SIB, BPT, and DLD test data were used in consensus diagnosis decisions. The diagnosis of dementia required a clinical and neurological examination showing deficits in 2 or more areas of cognitive functioning, and progressive worsening of cognitive performance compared to the potential participant's baseline functioning.

2.2.2. Data Preparation

Raw scores were used for all measures except for the Vineland-II domain-level scores (ABC, communication, daily living skills, socialization, and motor skills), which were only available as standardized scores. No subscales were removed for excessive missingness (>15% of data points missing across individuals). The DLD had the least amount of missing data (0.71% missing), followed by the Vineland-II (9.22% missing) and the SIB (9.93% missing). Missing data were imputed using chained random forests via the 'missRanger' R package [35]. Next, the DLD scores were inverted for consistent directionality with the other measures. All analyses were completed in R v 4.0.0 [36] and the significance level set to 0.05.

Aim 1: Principal Components Analysis

For the first aim, principal components analysis (PCA) was used to identify the number of components assessed by all individual items from the performance and informant procedures. The R package 'tidymodels' was used for all steps of the PCA analysis. The appropriateness of using PCA was evaluated using variable correlations, Bartlett's test, Kaiser-Meyer-Olkin, and determinants. In terms of correlations, variables should be only mildly intercorrelated and were examined using thresholds suggested by Field et al. [37] to have absolute correlations ranging from 0.3 to 0.9. Items with more than one occurrence for a correlation outside of the range were excluded from the PCA analysis. Only four variables needed to be excluded: Vineland-II ABC, Vineland-II social domain score, one item from the Vineland-II maladaptive behavior domain scale, and the SIB total score. The Kaiser-Meyer-Olkin factor adequacy was 0.95, above the 0.7 threshold. Bartlett's test of sphericity was significant ($X^2(325) = 5207.62$; $p < 0.001$). Finally, the determinant was below 0.00001. Together, these indicated that PCA was appropriate. Based on the scree plot of unrotated results, two components with eigenvalues > 1.0 were identified, accounting for 76% of the total variance in scores. Varimax rotation of loadings was then employed to enhance interpretability of identified components.

The dataset containing the two components scores and AD diagnostic status were then split into a training and test dataset. The training dataset was used to generate the logistic regression model. The model was assessed for multicollinearity and the assumption that independent variables are linearly related to the log odds. The performance of the generated model was assessed on the test dataset by evaluating the area under the curve (AUC), sensitivity, and specificity.

Aim 2: Classification and Regression Tree Analysis

For the second aim, classification and regression tree (CART) modeling was used to identify an optimal set of rules for classifying participants by diagnosis based only on item- and subscale-level data from the neurobehavioral battery. Again, R package 'tidymodels' was used in all steps of the CART analysis. First, training and test datasets were generated from the data. Then, the training dataset was used to generate a set of 10-fold cross-validation samples for model hyperparameter tuning. The best hyperparameters were selected based on the AUC. The CART model was first fit on the training dataset, then on the test dataset to assess performance.

Aim 3: Receiver Operating Characteristic Curves and Comparisons

For the third aim, to compare the relative utility of the PCA and CART models, the AUC of both models for classifying diagnosis were compared using the bootstrap test for comparing ROC curves (R routine 'roc.test' from the package 'pROC'). Sensitivity, specificity, and positive and negative predictive values were computed for the PCA and CART models.

3. Results

3.1. Sample Characteristics

A total of 141 participants were included in the current study. Just over half of the participants (n = 73; 51.77%) were diagnosed with probable AD. Full participant characteristics are provided in Table 1.

Table 1. Participant Characteristics.

Characteristic		No Dementia, N = 68	Probable AD, N = 73	Overall, N = 141
Sex				
	Female	36 (52.94%)	41 (56.16%)	77 (54.61%)
	Male	32 (47.06%)	32 (43.84%)	64 (45.39%)
Age (years)		38.11 (9.34)	52.68 (6.12)	45.66 (10.70)
Level of Intellectual Disability (estimated)				
	Mild	3 (4.41%)	14 (19.18%)	17 (12.06%)
	Moderate	36 (52.94%)	29 (39.73%)	65 (46.10%)
	Profound	28 (41.18%)	15 (20.55%)	43 (30.50%)
	Severe	1 (1.47%)	13 (17.81%)	14 (9.93%)
	Unknown	0 (0.00%)	2 (2.74%)	2 (1.42%)
Site				
	UCI	0 (0.00%)	53 (72.60%)	53 (37.59%)
	UKY	68 (100.00%)	20 (27.40%)	88 (62.41%)

n (%); mean (SD).

3.2. Aim 1: Principal Components Analysis Results

Results of the PCA are listed in Supplementary Table S1. The two components could not be easily labeled because they each contained items from communication, daily living skills, and cognitive domains. For the PCA method, logistic regression results demonstrated that higher scores on PC1 and PC2 were predictive of AD diagnosis. For each unit increase in PC1, the odds of a probable dementia diagnosis increased 2.54 times, and for each unit increase in PC2 the odds of a probable dementia diagnosis increased 3.73 times (Table 2).

Table 2. Logistic Regression.

Predictors		Odds Ratio	95% CI	p
(Intercept)		1.14	0.46–2.84	0.773
	PC1	2.54	1.69–3.81	<0.001
	PC2	3.73	1.62–8.60	0.002
Observations		106		
Tjur's R^2		0.786		

n (%); mean (SD).

3.3. Aim 2: Classification and Regression Tree Analysis Results

The CART analysis revealed that the DLD SCS and Vineland-II community subdomain raw scores best-classified dementia (Figure 1). A DLD SCS less than 35 and a Vineland community score less than 34 are indicative of AD dementia.

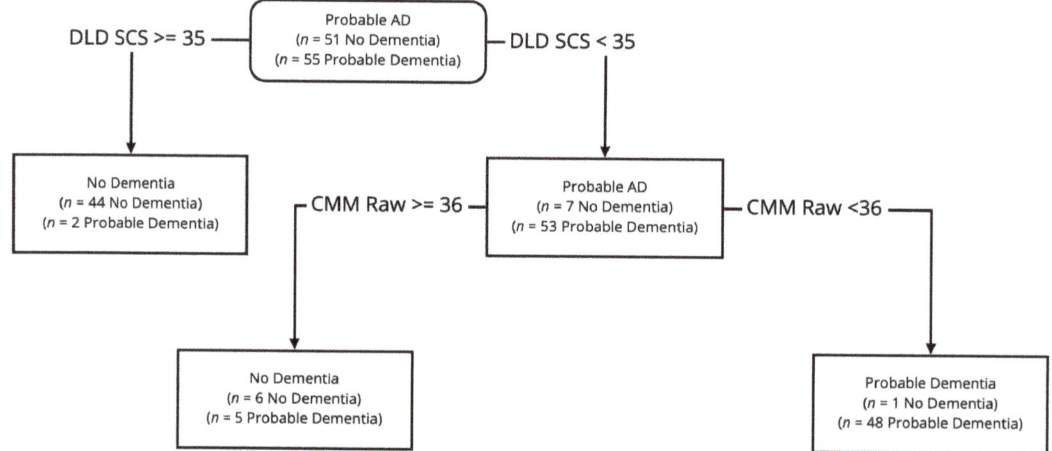

Figure 1. Results of CART analysis. *Note.* AD = Alzheimer's disease; DLD SCS = Dementia Questionnaire for People with Learning Disabilities (DLD) sum of cognitive scores raw score; CMM Raw = Vineland-II community subdomain raw score.

3.4. Aim 3: Comparison of PCA and CART Model Classification Utility

Comparing the PCA logistic regression and CART classification methods, there was no significant difference in AUC (D(65.196) = -0.57683; $p = 0.57$) (Figure 2). The PCA analysis resulted in an AUC of 0.87 while the CART model produced an AUC of 0.91. In terms of classification utility, the PCA model showed very good sensitivity (0.80) and good specificity (0.70), with high negative predictive value (0.824) and moderately high positive predictive value (0.667) at the combined sample base rate. The CART model demonstrated excellent sensitivity (1.00) and very good specificity (0.810), with excellent negative predictive value (1.00) and high positive predictive value (0.778) at the combined sample base rate.

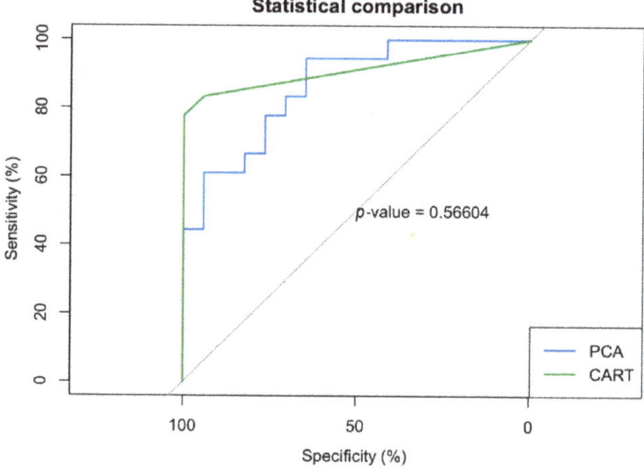

Figure 2. ROC curve comparison for PCA versus CART derived models. PCA area under the curve (AUC) = 0.87, and CART model AUC = 0.91.

4. Discussion

The present data indicate that for adults with DS, variability in orientation, language, memory, visuospatial skills, praxis, mood, and social participation is largely explained by two underlying principal components. These two components seemed to differentiate cognitive from practical function (i.e., the ability to answer questions vs. the ability to carry out everyday tasks). Additionally, the use of the two-component model to categorize participants with respect to AD dementia status showed high classification accuracy. These findings support the further distillation of the modest-sized battery into a "short form" that can be easily administered in a primary care setting. Additionally, it takes less than an hour to administer to an informant who knows the patient with DS well, and requires minimal office space and test stimuli.

Results of the CART analysis also demonstrated that a small subset of the original battery—the cognitive subscale of the DLD (SCS) and the community subscale of the Vineland-II—were just as effective in classifying AD dementia status. However, the CART model exhibited better negative predictive value, in that fewer participants with dementia were misclassified as non-demented compared to the principal components model. A short battery based on the CART model is also quicker to administer and can in most cases be completed in less than 30 min.

A key finding is that the two contributory measures are not direct, objective measures of cognitive performance completed by the patient. Instead, they are informant-based scores of the patient's observed changes in cognitive abilities (DLD-SCS) and self-management in community tasks (Vineland-II community). Unexpectedly, classification did not appreciably hinge on objective, performance-based neurocognitive measures. This highlights the critical component of informed caregiver ratings when screening for dementia in DS populations and provides some assurance that differential diagnosis of AD dementia is still possible when a patient's cognitive abilities cannot be directly assessed due to profound ID, limited cooperation, sensory impairments, or speech and language disorders.

Overall, the present data suggest that in clinical contexts with limited time and access to advanced training in test administration, the cognitive subscale of the DLD and community subscale of the Vineland-II, two widely available instruments, may suffice for screening and monitoring purposes. To be clear, we do not conclude that these two subscales constitute a comprehensive research or diagnostic battery, as definitive diagnosis should be based on longitudinal data. Nor is it the case that objective neurocognitive performance measures are redundant for diagnostic purposes. On the contrary, diagnostic criteria require objective neurocognitive assessment in order to make a firm diagnosis [33]. The present analysis was conducted for the specific aims of the study, namely identifying measures for resource-limited healthcare settings to encourage wide adoption of dementia screening among community DS practitioners. Prior efforts to use data reduction approaches to streamline a cognitive and behavioral battery for dementia in DS were focused primarily on developing a minimal comprehensive battery for research and specialty evaluation settings; thus, the resulting recommendations were not as relevant to primary care screening [38].

Furthermore, the present findings do not suggest that these two subscales represent an advancement in the early detection of AD dementia relative to more comprehensive test batteries. Instead, the benefit of adopting a minimal screening battery would enable more of the broader DS population to be evaluated, who may otherwise go unassessed. At the individual level, "early" detection is relative to the person's typical access to care, not the recommended standard of care. Given that nearly half of adults with DS do not receive regular screening for typical DS-associated health problems [11], it is reasonable to cast a wider net with "good enough" measures easily administered in primary care settings. Moreover, operating characteristics of the CART model align with a preference for high sensitivity (potential over-identification) over high specificity because the goal of screening is to provide support to this population.

Examination of the factor structure and of the item- and subscale-level operating characteristics in diagnostic batteries for dementia in DS is relatively new ground, and it is difficult to contextualize the present findings in the literature on the constructs measured. Broadly speaking, these data are in line with indications that adults with DS have reduced—but not absent—functional independence relative to other adults with intellectual disabilities [39], and dementia-related impairment in that domain may be captured by a community functioning measure such as the Vineland-II community subscale. Prior work using the Vineland-II to predict AD dementia in DS found that informant-rated receptive language skills, in addition to performance on a semantic verbal fluency task, were strong indicators of mild cognitive impairment in DS [40]. The present analysis instead examined individuals with and without AD dementia and found community management skills to be the most informative subscale of the Vineland-II. These findings are not contradictory, as in the present study it is likely that variability between participants cognitive functioning were captured by the DLD-SCS informant-based score, leaving more contextual community-based functioning to be best represented by the Vineland-II community subscale.

Beyond those discussed above, additional limitations of this study include the use of the SIB, DLD, and BPT along with the neurologic examination to determine consensus diagnosis. Our prior investigations have found that in 96% of cases, the final consensus diagnosis matched the neurologist's diagnosis that was formed independently of the SIB, BPT, and DLD scores. Still, discussion with the informant allows exposure to much of the same information captured by these instruments, and consideration of this information when forming a diagnosis is unavoidable. The eventual goal of both study cohorts is to substantiate consensus diagnoses with neuropathology at autopsy, allowing a more direct evaluation of the influence of potential criterion contamination. Additionally, the present study relied on informants who were very familiar with the participants with DS being rated, and in many cases, such a source of information cannot be found in practice.

Supplementary Materials: The following are available online at https://www.mdpi.com/article/10.3390/brainsci11091128/s1, Table S1: Principal Components Analysis Results.

Author Contributions: Conceptualization, J.P.H., L.M.K., K.L.V.P., C.L.H., E.D., E.H., I.T.L. and F.A.S.; methodology, J.P.H., L.M.K., K.L.V.P., C.L.H., E.H. and F.A.S.; software, K.L.V.P.; validation, K.L.V.P.; formal analysis, K.L.V.P.; investigation, J.P.H., L.M.K., K.L.V.P., C.L.H., E.D., E.H., I.T.L. and F.A.S.; resources, F.A.S. and E.H.; data curation, K.L.V.P. and E.D.; writing—original draft preparation, J.P.H., K.L.V.P. and L.M.K.; writing—review and editing, J.P.H., L.M.K., K.L.V.P., C.L.H., E.D., E.H., I.T.L. and F.A.S.; visualization, K.L.V.P.; supervision, J.P.H., L.M.K., K.L.V.P., C.L.H., E.D., E.H., I.T.L. and F.A.S.; project administration, J.P.H., L.M.K., K.L.V.P., C.L.H., E.D., E.H., I.T.L. and F.A.S.; funding acquisition, J.P.H., L.M.K., K.L.V.P., C.L.H., E.D., E.H., I.T.L. and F.A.S. All authors have read and agreed to the published version of the manuscript.

Funding: This research was funded by the National Institute on Aging, grant numbers P30 AG028383, 1T32AG057461, AG-21912, P50-16573, P30AG066519, and U01 AG051412, and the National Institute of Child Health and Human Development, grant number HDR01064993.

Institutional Review Board Statement: The study was conducted according to the guidelines of the Declaration of Helsinki and approved by the Institutional Review Boards of University of California, Irvine (HS#2002-2796; 08/07/2007) and University of Kentucky (#61423 2019; 09/03/2020).

Informed Consent Statement: Informed consent was obtained from all subjects involved in the study.

Data Availability Statement: The data presented in this study are openly available in OSFHome at [DOI 10.17605/OSF.IO/EK3YH].

Acknowledgments: We thank the participants and their families/caregivers for participating in this study. We are grateful to the University of California, Irvine Institute for Clinical and Translational Science (UL1 RR031985) for providing resources in support of this project. The authors would like to thank Stacey Brothers, Katie McCarty, Roberta Davis, Allison Caban-Holt, and Amelia Anderson-Mooney for their assistance with data collection. We also want to thank our team of neurologists: Gregory Jicha, Donita Lightner, and William Robertson for their clinical expertise.

Conflicts of Interest: The authors declare no conflict of interest. The funders had no role in the design of the study; in the collection, analyses, or interpretation of data; in the writing of the manuscript, or in the decision to publish the results.

References

1. Zigman, W.B.; Lott, I.T. Alzheimer's disease in Down syndrome: Neurobiology and risk. *Ment. Retard. Dev. Disabil. Res. Rev.* **2007**, *13*, 237–246. [CrossRef]
2. Pucharcos, C.; Fuentes, J.-J.; Casas, C.; de la Luna, S.; Alcantara, S.; Arbones, M.L.; Soriano, E.; Estivill, X.; Pritchard, M. Alu-splice cloning of human Intersectin (ITSN), a putative multivalent binding protein expressed in proliferating and differentiating neurons and overexpressed in Down syndrome. *Eur. J. Hum. Genet.* **1999**, *7*, 704–712. [CrossRef] [PubMed]
3. Wisniewski, K.; Wisniewski, H.; Wen, G. Occurrence of neuropathological changes and dementia of Alzheimer's disease in Down's syndrome. *Ann. Neurol.* **1985**, *17*, 278–282. [CrossRef] [PubMed]
4. Fuentes, J.J.; Genescà, L.; Kingsbury, T.J.; Cunningham, K.W.; Pérez-Riba, M.; Estivill, X.; Luna, S.D.L. DSCR1, overexpressed in Down syndrome, is an inhibitor of calcineurin-mediated signaling pathways. *Hum. Mol. Genet.* **2000**, *9*, 1681–1690. [CrossRef]
5. Zigman, W.B. Atypical aging in Down syndrome. *Dev. Disabil. Res. Rev.* **2013**, *18*, 51–67. [CrossRef]
6. Bush, A.; Beail, N. Risk factors for dementia in people with Down syndrome: Issues in assessment and diagnosis. *Am. J. Ment. Retard.* **2004**, *109*, 83–97. [CrossRef]
7. Maris, M.; Verhulst, S.; Wojciechowski, M.; Van de Heyning, P.; Boudewyns, A. Prevalence of obstructive sleep apnea in children with Down syndrome. *Sleep* **2016**, *39*, 699–704. [CrossRef]
8. Head, E.; Lott, I.T.; Wilcock, D.M.; Lemere, C.A. Aging in Down syndrome and the development of Alzheimer's disease neuropathology. *Curr. Alzheimer Res.* **2016**, *13*, 18–29. [CrossRef]
9. Lott, I.T.; Head, E. Dementia in Down syndrome: Unique insights for Alzheimer disease research. *Nat. Rev. Neurol.* **2019**, *15*, 135–147. [CrossRef] [PubMed]
10. Stoltzner, S.E.; Grenfell, T.J.; Mori, C.; Wisniewski, K.E.; Wisniewski, T.M.; Selkoe, D.J.; Lemere, C.A. Temporal accrual of complement proteins in amyloid plaques in Down's syndrome with Alzheimer's disease. *Am. J. Pathol.* **2000**, *156*, 489–499. [CrossRef]
11. Jensen, K.M.; Taylor, L.C.; Davis, M.M. Primary care for adults with Down syndrome: Adherence to preventive healthcare recommendations. *J. Intellect. Disabil. Res.* **2013**, *57*, 409–421. [CrossRef]
12. Cooley, W.C.; Graham, J.M., Jr. Common syndromes and management issues for primary care physicians: Down syndrome—An update and review for the primary pediatrician. *Clin. Pediatr.* **1991**, *30*, 233–253. [CrossRef]
13. Baum, R.A.; Nash, P.L.; Foster, J.; Spader, M.; Ratliff-Schaub, K.; Coury, D.L. Primary care of children and adolescents with Down syndrome: An update. *Curr. Probl. Pediatr. Adolesc. Health Care* **2008**, *38*, 241–261. [CrossRef] [PubMed]
14. Bull, M.J. Health supervision for children with Down syndrome. *Pediatrics* **2011**, *128*, 393–406. [CrossRef]
15. Landes, S.D.; Stevens, J.D.; Turk, M.A. Cause of death in adults with Down syndrome in the United States. *Disabil. Health J.* **2020**, *13*, 100947. [CrossRef]
16. Tsou, A.Y.; Bulova, P.; Capone, G.; Chicoine, B.; Gelaro, B.; Harville, T.O.; Martin, B.A.; McGuire, D.E.; McKelvey, K.D.; Peterson, M.; et al. Medical care of adults with Down syndrome: A clinical guideline. *JAMA* **2020**, *324*, 1543–1556. [CrossRef] [PubMed]
17. Steingass, K.J.; Chicoine, B.; McGuire, D.; Roizen, N.J. Developmental disabilities grown up: Down syndrome. *J. Dev. Behav. Pediatr.* **2011**, *32*, 548–558. [CrossRef]
18. Capone, G.T.; Chicoine, B.; Bulova, P.; Stephens, M.; Hart, S.; Crissman, B.; Videlefsky, A.; Myers, K.; Roizen, N.; Esbensen, A.; et al. Co-occurring medical conditions in adults with Down syndrome: A systematic review toward the development of health care guidelines. *Am. J. Med. Genet. A* **2018**, *176*, 116–133. [CrossRef]
19. Startin, C.M.; Hamburg, S.; Hithersay, R.; Al-Janabi, T.; Mok, K.Y.; Hardy, J.; Strydom, A.; Fisher, E.; Nizetic, D.; Tybulewicz, V.; et al. Cognitive markers of preclinical and prodromal Alzheimer's disease in Down syndrome. *Alzheimer's Dement.* **2019**, *15*, 245–257. [CrossRef] [PubMed]
20. Firth, N.C.; Startin, C.M.; Hithersay, R.; Hamburg, S.; Wijeratne, P.A.; Mok, K.Y.; Hardy, J.; Alexander, D.C.; Consortium, L.; Strydom, A. Aging related cognitive changes associated with Alzheimer's disease in Down syndrome. *Ann. Clin. Transl. Neurol.* **2018**, *5*, 741–751. [CrossRef] [PubMed]
21. Koehl, L.; Harp, J.; Van Pelt, K.L.; Head, E.; Schmitt, F.A. Longitudinal assessment of dementia measures in Down syndrome. *Alzheimer's Dement. Diagn. Assess. Dis. Monit.* **2020**, *12*, e12075.

22. Dalton, A.; Sano, M.; Aisen, P. Brief Praxis test: A primary outcome measure for treatment trial of Alzheimer disease in persons with Down syndrome. In *Multi-Centre Vitamine E Trial: ML Margallo-Lana et al. Project Proposal*; New York State Institute for basic Research in Developmental Disabilities: New York, NY, USA, 2001.
23. Panisset, M.; Roudier, M.; Saxton, J.; Boiler, F. Severe impairment battery: A neuropsychological test for severely demented patients. *Arch. Neurol.* **1994**, *51*, 41–45. [CrossRef]
24. Evenhuis, H.M. Further evaluation of the Dementia Questionnaire for Persons with Mental Retardation (DMR). *J. Intellect. Disabil. Res.* **1996**, *40*, 369–373. [CrossRef] [PubMed]
25. Sparrow, S.; Cicchetti, D.; Balla, D. *Vineland Adaptive Behavioral Scales*, 2nd ed.; Pearson: London, UK, 2006.
26. Winblad, B.; Kilander, L.; Eriksson, S.; Minthon, L.; Båtsman, S.; Wetterholm, A.-L.; Jansson-Blixt, C.; Haglund, A.; Severe Alzheimer's Disease Study Group. Donepezil in patients with severe Alzheimer's disease: Double-blind, parallel-group, placebo-controlled study. *Lancet* **2006**, *367*, 1057–1065. [CrossRef]
27. Lott, I.T.; Doran, E.; Nguyen, V.Q.; Tournay, A.; Head, E.; Gillen, D.L. Down syndrome and dementia: A randomized, controlled trial of antioxidant supplementation. *Am. J. Med. Genet. Part A* **2011**, *155*, 1939–1948. [CrossRef] [PubMed]
28. Sano, M.; Aisen, P.S.; Andrews, H.F.; Tsai, W.-Y.; Lai, F.; Dalton, A.J. Vitamin E in aging persons with Down syndrome: A randomized, placebo-controlled clinical trial. *Neurology* **2016**, *86*, 2071–2076. [CrossRef]
29. Schmitt, F.A.; Ashford, J.W.; Ferris, S.; Mackell, J.; Saxton, J.; Schneider, L.; Clark, C.; Ernesto, C.; Schafer, K.; Thal, L. Severe impairment battery: A potential measure for AD clinical trials. In *Alzheimer Disease*; Springer: Berlin/Heidelberg, Germany, 1997; pp. 419–423.
30. Graves, R.J.; Zlomke, K.; Graff, J.C.; Hall, H.R. Adaptive behavior of adults with Down syndrome and their health-related quality of life. *Adv. Neurodev. Disord.* **2020**, *4*, 27–35. [CrossRef]
31. Farmer, C.; Adedipe, D.; Bal, V.; Chlebowski, C.; Thurm, A. Concordance of the Vineland Adaptive Behavior Scales, second and third editions. *J. Intellect. Disabil. Res.* **2020**, *64*, 18–26. [CrossRef]
32. Gangadharan, S.K.; Jesu, A.J.M. Dementia in people with intellectual disability. In *The Frith Prescribing Guidelines for People with Intellectual Disability*, 3rd ed.; John Wiley & Sons Inc: Chichester, UK, 2015; pp. 63–75.
33. McKhann, G.M.; Knopman, D.S.; Chertkow, H.; Hyman, B.T.; Jack, C.R., Jr.; Kawas, C.H.; Klunk, W.E.; Koroshetz, W.J.; Manly, J.J.; Mayeux, R.; et al. The diagnosis of dementia due to Alzheimer's disease: Recommendations from the National Institute on Aging-Alzheimer's Association workgroups on diagnostic guidelines for Alzheimer's disease. *Alzheimer's Dement.* **2011**, *7*, 263–269. [CrossRef]
34. Strydom, A.; Livingston, G.; King, M.; Hassiotis, A. Prevalence of dementia in intellectual disability using different diagnostic criteria. *Br. J. Psychiatry* **2007**, *191*, 150–157. [CrossRef]
35. Mayer, M. missRanger: Fast Imputation of Missing Values. 2021. Available online: http://130.208.58.81/web/packages/missRanger/missRanger.pdf (accessed on 21 August 2021).
36. R Core Team. *R: A Language and Environment for Statistical Computing*; R Core Team: Vienna, Austria, 2013.
37. Field, A.; Miles, J.; Field, Z. *Discovering Statistics Using R*; Sage: New York, NY, USA, 2012.
38. Aschenbrenner, A.J.; Baksh, R.A.; Benejam, B.; Beresford-Webb, J.A.; Coppus, A.; Fortea, J.; Handen, B.L.; Hartley, S.; Head, E.; Jaeger, J.; et al. Markers of early changes in cognition across cohorts of adults with Down syndrome at risk of Alzheimer's disease. *Alzheimer's Dement. Diagn. Assess. Dis. Monit.* **2021**, *13*, e12184.
39. Chapman, R.S.; Hesketh, L.J. Behavioral phenotype of individuals with Down syndrome. *Ment. Retard. Dev. Disabil. Res. Rev.* **2000**, *6*, 84–95. [CrossRef]
40. Pulsifer, M.B.; Evans, C.L.; Hom, C.; Krinsky-McHale, S.J.; Silverman, W.; Lai, F.; Lott, I.; Schupf, N.; Wen, J.; Rosas, H.D. Language skills as a predictor of cognitive decline in adults with Down syndrome. *Alzheimer's Dement. Diagn. Assess. Dis. Monit.* **2020**, *12*, e12080. [CrossRef] [PubMed]

Article

Cognitive Function during the Prodromal Stage of Alzheimer's Disease in Down Syndrome: Comparing Models

Christy L. Hom [1,*], Katharine A. Kirby [2], Joni Ricks-Oddie [2,3], David B. Keator [1], Sharon J. Krinsky-McHale [4], Margaret B. Pulsifer [5], Herminia Diana Rosas [6], Florence Lai [6], Nicole Schupf [7,8], Ira T. Lott [9] and Wayne Silverman [9]

1 Department of Psychiatry and Human Behavior, University of California, Orange, CA 92868, USA; dbkeator@uci.edu
2 Center for Statistical Consulting, University of California, Irvine, CA 92697, USA; kathark@uci.edu (K.A.K.); jricksod@uci.edu (J.R.-O.)
3 Institute for Clinical and Translational Sciences, University of California, Irvine, CA 92697, USA
4 New York State Institute for Basic Research in Developmental Disabilities, Staten Island, NY 10314, USA; sharon.krinsky-mchale@opwdd.ny.gov
5 Department of Psychiatry, Massachusetts General Hospital, Harvard Medical School, Boston, MA 02114, USA; mpulsifer@mgh.harvard.edu
6 Department of Neurology, Massachusetts General Hospital, Harvard Medical School, Boston, MA 02114, USA; rosas@helix.mgh.harvard.edu (H.D.R.); flai@partners.org (F.L.)
7 Department of Neurology, College of Physicians and Surgeons, Columbia University, New York, NY 10027, USA; ns24@cumc.columbia.edu
8 Department of Epidemiology, School of Public Health, Columbia University, New York, NY 10027, USA
9 Department of Pediatrics, University of California, Orange, CA 92868, USA; itlott@uci.edu (I.T.L.); wsilverm@uci.edu (W.S.)
* Correspondence: homc@uci.edu; Tel.: +1-714-456-2927

Citation: Hom, C.L.; Kirby, K.A.; Ricks-Oddie, J.; Keator, D.B.; Krinsky-McHale, S.J.; Pulsifer, M.B.; Rosas, H.D.; Lai, F.; Schupf, N.; Lott, I.T.; et al. Cognitive Function during the Prodromal Stage of Alzheimer's Disease in Down Syndrome: Comparing Models. *Brain Sci.* **2021**, *11*, 1220. https://doi.org/10.3390/brainsci11091220

Academic Editor: Sien Hu

Received: 13 July 2021
Accepted: 14 September 2021
Published: 16 September 2021

Publisher's Note: MDPI stays neutral with regard to jurisdictional claims in published maps and institutional affiliations.

Abstract: Accurate identification of the prodromal stage of Alzheimer's disease (AD), known as mild cognitive impairment (MCI), in adults with Down syndrome (MCI-DS) has been challenging because there are no established diagnostic criteria that can be applied for people with lifelong intellectual disabilities (ID). As such, the sequence of cognitive decline in adults with DS has been difficult to ascertain, and it is possible that domain constructs characterizing cognitive function in neurotypical adults do not generalize to this high-risk population. The present study examined associations among multiple measures of cognitive function in adults with DS, either prior to or during the prodromal stage of AD to determine, through multiple statistical techniques, the measures that reflected the same underlying domains of processing. Participants included 144 adults with DS 40–82 years of age, all enrolled in a larger, multidisciplinary study examining biomarkers of AD in adults with DS. All participants had mild or moderate lifelong intellectual disabilities. Overall AD-related clinical status was rated for each individual during a personalized consensus conference that considered performance as well as health status, with 103 participants considered cognitively stable (CS) and 41 to have MCI-DS. Analyses of 17 variables derived from 10 tests of cognition indicated that performance reflected three underlying factors: language/executive function, memory, and visuomotor. All three domain composite scores significantly predicted MCI-DS status. Based upon path modeling, the language/executive function composite score was the most affected by prodromal AD. However, based upon structural equation modeling, tests assessing the latent construct of memory were the most impacted, followed by those assessing visuomotor, and then those assessing language/executive function. Our study provides clear evidence that cognitive functioning in older adults with DS can be characterized at the cognitive domain level, but the statistical methods selected and the inclusion or exclusion of certain covariates may lead to different conclusions. Best practice requires investigators to understand the internal structure of their variables and to provide evidence that their variables assess their intended constructs.

Keywords: mild cognitive impairment; Alzheimer's disease; Down syndrome; cognitive decline; neuropsychological tests; cognitive function

1. Introduction

Early detection of Alzheimer's disease (AD) is of paramount importance to enhance efficacy of clinical intervention and to improve understanding of AD progression in people with Down syndrome (DS). Due to the triplication of the amyloid precursor protein gene on chromosome 21, people with DS have life-long overproduction of amyloid-β, and the earlier production of amyloid- β in this population, detectable in imaging studies as early as 14 years of age, contributes to increased AD risk [1,2]. That is, adults with DS are likely to experience earlier AD-related cognitive declines than their counterparts in the neurotypical population [3].

Complicating early detection of AD progression and diagnostic accuracy is the large variability in baseline cognitive ability among people with DS, ranging from borderline intellectual functioning to profound impairment [2,4,5]. Diagnosis of any form of intellectual disability (ID) requires an intelligence quotient (IQ) that is two or more standard deviations (SD) below the population mean, and the range in IQ within the population with DS can be more than 50 points [6], equivalent to over 3 SDs. In the neurotypical population, the prodromal stage of dementia (i.e., mild cognitive impairment; MCI) is typically diagnosed based upon decline, together with performance that is 1.5 SDs or more below the population mean on cognitive tests [7,8]). However, this criterion cannot be applied to the population with DS since the vast majority of affected individuals have functioned more than 1.5 SDs below the mean since childhood [2,3]. The lack of data on normative age-related declines in cognition among healthy adults with DS also makes it difficult to differentiate between expected age-related declines and MCI [9]. A 50-year study by Carr and Collins [10] is one of the few longitudinal investigations in people with DS that has been able to isolate age-related from disease-related changes, and they found that in the absence of dementia, people with DS experienced significant age-related changes in some cognitive domains (e.g., memory and non-verbal IQ) but not others (e.g., receptive and expressive language skills).

The lack of agreed-upon diagnostic criteria for dementia specific to people with ID, in the literature or in practice, further complicates the diagnostic process [11], which is why Basten et al. [12] argued that recognition and treatment of the neurodegenerative components of the syndrome are the greatest unmet therapeutic needs. The lack of specifics regarding the subtle cognitive changes that characterize prodromal AD not only affects accuracy of diagnosis and the timing of pharmacological interventions for the population with DS, but it also contributes to their being overlooked for inclusion in AD randomized controlled trials [13].

Numerous investigations have been undertaken to characterize the cognitive changes associated with AD in persons with DS. Most of these investigations have been unsuccessful in identifying consistent or clear differences between those in the preclinical and prodromal stages of AD [13,14]. Some found changes in episodic memory to be prominent during the prodromal and early stage of AD in DS [15], whereas others found changes in personality/behavior and executive function to be more prominent than deterioration in episodic memory in persons with severe to mild ID [16–21]. Still others found reduced language skills [22–24] and adaptive functioning [25,26] to be the earliest indicators. After a systematic review of the DS-AD literature, Lautarescu et al. [27] concluded that some of the variability in presentation during the early stages of AD in the population with DS may be due to differences in premorbid intellectual capacity and each individual's ability to compensate for newly acquired deficits.

The present study explores whether cognitive deficits characterizing the prodromal stage of AD in individuals with DS reflect the same underlying domains when severity of intellectual disability ranges from "moderate" to "mild". Specifically, we examine tasks requiring executive function, language, memory, or visuomotor coordination, areas previously found to be particularly affected by dementia in adults with DS [28]. We also examine age-related declines in this sample that is at high risk for AD neuropathology as well as compare different methods for modeling the cognitive decline associated with AD progression. Due to the exploratory nature of these analyses, we did not specify a priori a hierarchy of deterioration among cognitive domains.

2. Materials and Methods

2.1. Participants and Procedure

All procedures were reviewed and approved by the Institutional Review Boards at participating institutions (New York State Institute for Basic Research in Developmental Disabilities, Columbia University Irving Medical Center, Massachusetts General Hospital, the University of California at Irvine, The Johns Hopkins University Schools of Medicine and Public Health, and the University of North Texas Health Science Center). Informed consent was obtained from participants or their legally authorized representatives along with participant assent. The current analyses included data from 144 adults with DS, ranging from 40 to 81 years of age, who were enrolled in a larger, multidisciplinary program of research focused on biomarkers of AD in adults with DS. Only the participants assessed in New York, New York ($n = 43$); Boston, Massachusetts ($n = 56$); and Irvine, California ($n = 46$) were administered the full neuropsychological battery that is reported in this study. Therefore, participants assessed at other sites could not be included in the current set of analyses. Inclusion criteria included (1) age ≥ 40, (2) estimated preexisting IQ > 30, (3) absence of significant sensory or motor impairments, and (4) willingness to provide a routine blood sample for studies of fluid-based biomarkers of AD.

Participants received a comprehensive evaluation at study baseline that included (a) a review of medical records; (b) physical and neurological evaluations; (c) interviews with knowledgeable informants focused on cognitive and functional abilities, health-related condition and medical history, and neuropsychiatric concerns; and (d) direct one-on-one testing with a core battery developed specifically for assessing dementia status in adults with intellectual disabilities. The full direct testing battery required approximately 1.5 to 2 h to complete. None of the participants in the larger, multidisciplinary program of research with severe ID had valid scores on all 22 measures; therefore, this subgroup was not included in the current study.

2.2. Consensus Disease Status

Following the comprehensive evaluation, each participant's AD-related disease status was determined through a Consensus Review Conference (see [29]) that included program investigators at the respective enrollment site, senior staff members, and research staff who had direct contact with the participant under consideration. Disease status was classified into the following categories: (a) Cognitively Stable (CS), indicating with reasonable certainty that AD-related impairment was absent (although allowing for declines normally expected to accompany aging, per se); (b) MCI-DS, indicating that there was some indication of cognitive and/or functional decline beyond what would be expected with aging, per se, but of insufficient severity to suggest frank dementia; (c) Possible Dementia, indicating that some signs and symptoms of dementia were present but were not judged to be totally convincing; (d) Definite Dementia, indicating with high confidence that dementia was present; and (e) Uncertain (due to complications), indicating that evidence of clinically significant declines were present but might be caused by some other substantial concern, usually a medical condition unrelated to a dementing disorder or a significant life event (e.g., severe sensory loss, poorly resolved hip fracture, death of a loved one). For the current study, only data from the CS and MCI-DS groups were analyzed.

2.3. Measures

The core neuropsychological battery consisted of instruments previously demonstrated to be valid for use with adults having DS and that covered the spectrum of cognitive domains expected to be affected by clinical progression of AD, including during its prodromal stage. Twenty-two measures from the following 13 instruments were hypothesized a priori to measure four cognitive domains: (1) executive function, (2) language, (3) memory, and (4) visuomotor:

1. The Beery Buktenica Developmental Test of Visual-Motor Integration—long form (VMI; [30]) assesses visual-motor integration skills. The total raw score is used.
2. The Block Design subtest from the Wechsler Intelligence Scale for Children, 4th Edition [31] supplemented with less complex items from the original Down Syndrome Mental Status Examination (DSMSE) [32] assesses visual–spatial reasoning and visual–motor dexterity. The total raw score is used.
3. The Boston Naming Test [33] measures confrontational picture-naming abilities. The measure of performance is total correct, with or without a semantic cue.
4. An adaptation of the Category Fluency Test [34] with slightly liberalized scoring measures semantic fluency. Participants are asked to generate as many words as they can within one of three categories (food, animals, clothing) in 20 s. The measure of performance is the total number of words generated in two categories, excluding repetitions.
5. The Cats and Dogs Task [20] assesses response inhibition using a Stroop test paradigm [35]. The measure of performance is the amount of time used to name all of the animals as printed on the sheet (Naming condition) subtracted from the amount of time used to state the opposite animal (Switch condition). The number of errors on the Naming and Switch conditions are also recorded.
6. The Cued Recall Test [36] assesses verbal learning and memory. The measure of performance is total number correct across three test trials.
7. An enhanced version of the DSMSE [32] that expands the number of items included in tests of short-term memory (from 3 to 9 objects) assesses several different abilities. Three subscale scores are used: Language, Memory, and Visual Spatial.
8. A simplified version of the Modified Mini Mental Status Evaluation (mMMSE) [37,38] assesses several different abilities. Four subscales are used: Anomia, Concentration, Fine Motor, and combined Orientation (Person, Place, and Time).
9. The Rapid Assessment of Developmental Disabilities, Second Edition (RADD-2) [39] is a battery of items from commonly used tests, including the Wechsler Intelligence Scale for Children, 4th Edition; Hawaii Early Learning Profile [40]; and Merrill-Palmer-Revised Scales of Early Childhood [41]. The Digit Span Forward, Expressive Language, Hand Movements, Imitation, Receptive Language, and Similarities subscales are used. Since the Hand Movements and Imitations subscales both assess motor coordination, they were combined into one score for our analyses, labeled "sensorimotor".
10. Purdue Pegboard [42] assesses hand dexterity. The measure of performance is the number of pairs completed with both hands simultaneously within one minute.
11. The Rivermead Behavioural Memory Test [43] assesses visual memory and recognition. The measures of performance are the number correct during the identification trial and the number correct minus the number incorrect during the recognition trial.
12. A modified version of the Selective Reminding Test (SRT), now requiring free recall of 8 items over 3 trials (see [15,44]), assesses verbal learning and memory. The measure of performance is total correct over 3 trials.
13. The Tinetti Balance and Gait Assessment Tool [45] assesses gait and balance on a three-point ordinal scale with a range of 0 to 2. A score of 0 represents the most impairment, whereas a score of 2 represents independence. We only administered the Gait tool, and that score is presented.

2.4. Statistical Analyses

Statistical analyses were performed using Stata version 16. Age differences between the CS and MCI-DS groups were evaluated through a Mann–Whitney test because age was not normally distributed. All other demographic and health comparisons were evaluated through Pearson chi-square tests. Before evaluating how well our measures predicted MCI-DS status, we examined the underlying factor structure of all the measures, stratified by premorbid ID level, using an exploratory factor analysis (EFA) with oblique promax rotation to take account of the correlated nature of the domains and increase the interpretability of the factor pattern matrix. Age-related cognitive deficits were examined by performing separate linear regression models for each test, first for the CS group, then for the MCI-DS group. Domain composite scores were created by using the sum of all rescaled test scores that were hypothesized to assess that single domain. Test scores were rescaled using the Proportion of Maximum Scoring [46] (POMS), which uses a 0–1 range to show the magnitudes of associations among variables without changing the shape of the distribution. Path modeling was used to assess the sensitivity of domain composites in identifying MCI-DS, adjusting for sex and premorbid ID. Structural equation modeling (SEM) was used to assess the association between MCI-DS and each latent variable, adjusting for sex and premorbid ID. Domain labels were based upon the results of EFA and an a priori factor structure. Model fit of the path and final SEM models were compared using multiple fit indices, including the chi-square goodness of fit, comparative fit index (CFI), Tucker–Lewis Index (TLI), and root mean square error of approximation (RMSEA). An RMSEA below 0.08 and a CFI and TLI above 0.90 are considered to indicate acceptable fit [47,48].

3. Results

3.1. Demographics

In the full sample, 59.0% were male, 53.5% had premorbid ID in the mild range, and 46.5% in the moderate ID range. Most participants were white (86.1%), followed by Hispanic (6.9%), Asian (4.2%), Black (2.1%), and American Indian (0.7%). In terms of chromosomal diagnosis, 84.0% had full trisomy 21, 4.9% were mosaic, 3.5% had translocation DS, and 7.7% were unknown. Regarding AD status, 71.5% were CS and 28.5% were MCI-DS. The MCI-DS group was significantly older, $z = -3.39$, $p = 0.0001$, and were reported to have more co-occurring health problems than the CS group, but the rates of chronic medical conditions did not differ significantly ($ps = 0.126$ to 0.894, see Table 1).

Table 1. Demographic and health comorbidities by AD status.

Condition	CS (n = 103)	MCI-DS (n = 41)	Mann–Whitney/χ^2 Statistic	*p*-Value
Age	M = 48.65, SD = 6.27	M = 52.88, SD = 6.72	−3.39	0.0001
Sex	Male (56.31%)	Male (65.85%)	1.10	0.293
Premorbid ID	Mild (58.25%) Moderate (41.75%)	Mild (41.46%) Moderate (58.54%)	3.32	0.068
Depression	28.16%	29.27%	0.02	0.894
Diabetes	6.80%	9.76%	0.75	0.688
Hearing	Corrected (15.53%) Impaired (16.50%)	Corrected (19.51%) Impaired (19.51%)	0.65	0.723
Hypertension	7.46%	11.76%	0.51	0.473
Obstructive sleep apnea	30.10%	41.46%	4.14	0.126
Seizures	11.65%	9.76%	0.94	0.625
Vision	Corrected (56.41%) Impaired (17.48%)	Corrected (56.10%) Impaired (29.27%)	3.67	0.159

3.2. Test Outcomes and Cognitive Domains

Table 2 presents the means and standard deviations for each test and their hypothesized cognitive domain. Outcomes on all measures were in the expected direction (CS > MCI-DS).

Table 2. Means and standard deviations of raw test scores by AD status and their hypothesized cognitive domain.

Variable (Range of Scores)	Domain	CS (n = 103)	MCI-DS (n = 41)	Mann–Whitney U (p-Value)
Block Design (0–54)	Visuomotor	23.79 (10.49)	17.37 (12.78)	3.11 (0.002)
Boston Naming (0–27)	Language	15.88 (5.51)	13.23 (6.91)	2.04 (0.041)
Category Fluency (0–17)	EF	8.23 (3.18)	6.82 (3.68)	2.21 (0.027)
Cats and Dogs Switch (−17.00–61.80) [†]	EF	9.80 (11.22)	5.05 (11.21)	1.82 (0.069)
Cued Recall (3–35)	Memory	28.61 (6.83)	21.38 (9.22)	4.43 (0.0001)
DSMSE Language (3–52)	Language	37.10 (0.07)	30.80 (9.63)	3.28 (0.001)
DSMSE Memory (0–23)	Memory	14.08 (4.68)	9.80 (4.57)	4.62 (0.0001)
DSMSE Visual Spatial (2–8)	Visuomotor	6.18 (1.09)	5.59 (1.01)	2.98 (0.003)
mMMSE-DS Anomia (4–20)	Language	18.18 (2.20)	16.49 (4.19)	1.55 (0.122)
mMMSE-DS Concentration (0–6)	EF	3.69 (2.10)	2.39 (2.14)	3.15 (0.002)
mMMSE-DS Fine Motor (1–10)	Visuomotor	8.04 (1.29)	7.05 (2.28)	2.58 (0.010)
mMMSE-DS Orientation (5–30)	EF	25.46 (5.33)	20.87 (6.41)	4.75 (0.0001)
Purdue Pegboard Both Hands (0–8)	Visuomotor	2.62 (1.82)	1.51 (1.71)	3.14 (0.002)
RADD-2 Digit Span Forward (0–8)	EF	4.02 (1.66)	2.97 (1.80)	2.84 (0.005)
RADD-2 Expressive Lang. (0–16)	Language	11.24 (3.92)	8.80 (4.01)	3.34 (0.001)
RADD-2 Receptive Lang. (0–12)	Language	7.56 (2.51)	6.80 (2.99)	1.08 (0.279)
RADD-2 Sensorimotor (0–7)	Visuomotor	6.70 (0.50)	5.85 (1.50)	3.87 (0.0001)
RADD-2 Similarities (0–4)	Language	2.51 (1.49)	1.56 (1.47)	3.40 (0.001)
Rivermead Recognition (0–10)	Memory	5.42 (3.77)	2.21 (3.40)	3.90 (0.0001)
Selective Reminding Test (1–24)	Memory	15.50 (5.62)	9.18 (4.32)	5.70 (0.0001)
Tinetti Gait (4–12)	Visuomotor	10.75 (1.62)	10.37 (1.88)	1.47 (0.143)
VMI (1–25)	Visuomotor	15.45 (3.22)	13.98 (3.40)	1.75 (0.081)

[†] Cats and Dogs is measured in seconds; all other scores are measured in points (number correct). EF = executive function.

3.3. Underlying Structure

A series of exploratory factor analyses (EFAs) were performed to examine how well the cognitive variables loaded onto the four hypothesized domains. An initial set of analyses considered 22 scores, but results revealed problems with the following five variables: (1) the Tinetti Gait measure did not load on the first four factors and had a uniqueness of 0.93, meaning that much of the information contained in the variable was not predicted by any factor; (2) Cats and Dogs performance time did not load on the first four factors and had a uniqueness of 0.96, whereas the error score loaded on the factor describing memory and 28% of the sample had less than 50% accuracy; (3) mMMSE Orientation and mMMSE Concentration loaded on the same factor in every iteration of EFA, so the Orientation measure was dropped due to consistently lower loadings than the Concentration measure and a narrower range of abilities tested; (4) mMMSE Fine Motor loaded onto the factor that assessed language and executive function instead of visuomotor because it required participants to know how to write numbers and letters in order; and (5) Category Fluency loaded onto multiple factors and the loadings were weaker in relation to the other tests within the same factor (0.32 to 0.36). These five variables were therefore dropped from further analyses.

The results of the EFA with promax rotation for the 17 remaining variables are summarized in Table 3. We retained three factors since the third factor had an eigenvalue close to 1 (0.95). EFAs were also performed separately for the two levels of premorbid ID. Both analyses yielded similar results, with three factors accounting for 91.65% of the total variance in the mild ID group and 90.75% of the total variance in the moderate ID group. Furthermore, each measure loaded onto the same factor, regardless of premorbid level of ID. The three factors consistently identified by each EFA were (1) language/executive function, (2) memory, and (3) visuomotor, confirming three out of four a priori hypothesized domains. These three factors explained 97.55% of the total variance in scores among the full sample across all levels of ID.

Table 3. Factor structure of the 17 retained cognitive variables across all levels of ID (*n* = 128).

	1	2	3
	Language/Executive Function	Visuomotor	Memory
EXECUTIVE FUNCTION			
mMMSE Concentration	0.67		
RADD-2 Digit Span Forward	0.82		
LANGUAGE			
Boston Naming	0.91		
DSMSE Language	0.98		
mMMSE Anomia	0.70		
RADD-2 Expressive Language	0.79		
RADD-2 Receptive Language	0.61		
RADD-2 Similarities	0.80		
MEMORY			
Cued Recall			0.80
DSMSE Memory			0.80
Rivermead Recognition			0.62
Selective Reminding Test			0.66
VISUOMOTOR			
Block Design		0.80	
DSMSE Visual Spatial		0.83	
Purdue Pegboard		0.49	
RADD-2 Sensorimotor		0.57	
VMI		0.54	
Percentage of variance	74.13	16.51	6.91

Blanks represent absolute loading < 0.30.

Although our executive function measures all required language skills to a certain extent, we did not expect them to be as highly correlated with the language measures (r^2 = 0.53 to 0.72, $p < 0.001$) as they were with each other (r^2 = 0.67, $p < 0.001$). As a result, tests from both domains were combined into a single factor, labeled "language/executive function." As mentioned above, two of the other tests that were hypothesized to measure executive function (Category Fluency and Cats and Dogs) were dropped from all additional analyses because they lacked clear association with a single factor that had an eigenvalue of ≥1 and/or had relatively weak loadings (defined a priori as anything < 0.30).

3.4. Effects of Aging

Linear regression models were performed to evaluate the effects of aging independent of disease status. Scatterplots and the fitted regression line for each cognitive test are presented as Figure S1. There was a significant main effect for aging on nine of the 17 measures, controlling for disease status. The three memory and two language measures that were affected by aging were DSMSE Language, DSMSE Memory, Purdue Pegboard, RADD-2 Expressive Language, RADD-2 Receptive Language, Rivermead Recognition, and SRT (Fs(3, 130) = 3.33 to 25.48, ts = −3.62 to −2.01, *ps* = 0.02 to 0.0001). Since there was only one significant interaction between age and disease status, RADD-2 Sensorimotor (F(3, 130) = 5.04, t = −5.08, *p* = 0.002), we decided not to include age as a covariate in all subsequent analyses in order to minimize the likelihood of a type II error.

3.5. Path Model

Path analysis was used to examine the magnitude and significance of the three domain composites, adjusting for sex and premorbid ID level. Every domain composite was significantly related to MCI-DS status (see Table 4, but the path model only explained about half of the variance between CS and MCI-DS (R^2 = 0.49) and had poor model fit, as the RMSEA was 0.49 (much greater than the 0.08 limit), CFI was 0.51 (not close to the 0.90 threshold), and TLI was −0.94 (far from the 0.90 cutoff).

Table 4. Path model comparing CS to MCI-DS adjusted for sex and premorbid ID level (*n* = 144).

Domain Score	β	SE	Z	P	95% CI Lower	Upper
Language/Executive Function	−0.97	0.29	−3.34	0.001	1.54	−0.40
Memory	−0.89	0.15	−6.04	0.0001	−1.17	−0.60
Visuomotor	−0.53	0.12	−4.23	0.0001	−0.77	−0.28

MCI-DS status was associated with lower scores on every domain composite, with the language/EF composite being the most affected, followed by the memory composite, then the visuomotor composite.

3.6. Structural Equation Model

Figure 1 displays the relationship between the latent, measurement, and exogenous variables (sex, premorbid ID, and MCI-DS status, respectively). The 17 cognitive measures and their latent variables predicted MCI-DS status substantially better than the composite scores used in the path model. The SEM model had good model fitness since RMSEA = 0.08, CFI = 0.91, and TLI = 0.90.

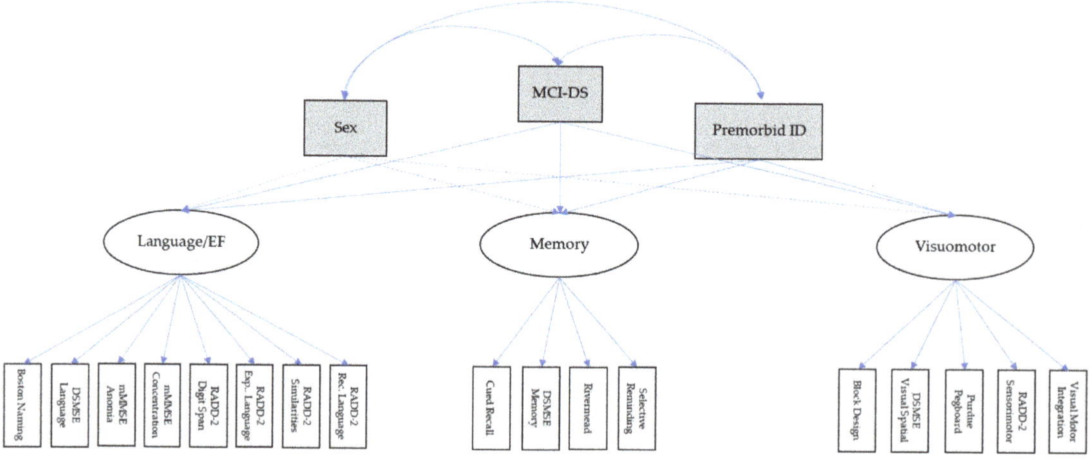

Figure 1. Structural equation model of how MCI-DS affected performance on the 17 cognitive measures and 3 latent variables. Solid lines indicate paths that were statistically significant, *ps* = 0.001. Ovals represent the endogenous variables with their indicator variables represented by white rectangles. CFI = Comparative Fit Index, CI = confidence interval, DSMSE = Down Syndrome Mental Status Examination, EF = executive function, Exp = expressive, ID = intellectual disability, mMMSE = Modified Mini-Mental Status Exam, MCI-DS = Mild Cognitive Impairment-Down syndrome, RADD-2 = Rapid Assessment of Developmental Disabilities-2nd Edition, Rec = receptive, RMSEA = root mean square error of approximation, Model Fit Indices: Chi-square = 299.20, CFI = 0.91, RMSEA = 0.08, 90% CI = 0.07–0.09.

Sex was not related to outcomes on any of the cognitive measures. MCI-DS status was associated with greater memory impairment than premorbid ID level was, whereas premorbid ID level was associated with greater language/EF impairment than MCI-DS status was. MCI-DS status and premorbid ID had similar effects on visuomotor performance.

As shown in Table 5, the latent construct of memory was a stronger indicator of MCI-DS status than the latent construct of visuomotor, which was a stronger indictor than language/EF. Although the path and SEM models indicated the same directionality of domain impairments for those with MCI-DS, the magnitude and ranking were different. (The path model identified the language/EF composite to be a stronger indicator of MCI-DS than the memory and visuomotor composites.) In short, giving all tests within the same domain differential weights as in the SEM can lead to considerably different conclusions than when all tests within the same domain are given equal weight as in the path analysis.

Table 5. The relationship between sex, premorbid ID, MCI-DS, and cognitive functioning.

Cognitive Domain	Predictor Variable	Estimate	SE	CR	Standardized Regression Weight	*p*-Value
		Structural				
Language/EF	Sex	−0.010	0.075	−0.13	−0.01	0.893
Language/EF	PID	−0.413	0.067	−6.15	−0.41	<0.001
Language/EF	MCI-DS	−0.238	0.074	−3.23	−0.24	0.001
Memory	Sex	0.008	0.076	0.11	0.01	0.916
Memory	PID	−0.241	0.075	−3.21	−0.24	0.001
Memory	MCI-DS	−0.471	0.67	−7.07	−0.47	<0.001
Visuomotor	Sex	−0.093	0.081	−1.14	−0.09	0.255
Visuomotor	PID	−0.302	0.077	−3.91	−0.30	<0.001
Visuomotor	MCI-DS	−0.303	0.78	−3.90	−0.30	<0.001
		Measurement				
Language/EF	Boston Naming	0.879	0.023	38.18	0.88	<0.001
Language/EF	DSMSE Language	0.878	0.023	37.81	0.88	<0.001
Language/EF	mMMSE Anomia	0.742	0.041	18.14	0.74	<0.001
Language/EF	mMMSE Concentration	0.724	0.043	16.53	0.72	<0.001
Language/EF	RADD-2 Digit Span Forward	0.782	0.036	21.93	0.78	<0.001
Language/EF	RADD-2 Expressive Language	0.874	0.023	37.52	0.87	<.001
Language/EF	RADD-2 Receptive Language	0.683	0.047	16.53	0.68	<0.001
Language/EF	RADD-2 Similarities	0.760	0.038	20.05	0.76	<0.001
Memory	Cued Recall	0.673	0.053	12.65	0.67	<0.001
Memory	DSMSE Memory	0.858	0.032	27.09	0.86	<0.001
Memory	Rivermead Recognition	0.702	0.051	13.87	0.70	<0.001
Memory	Selective Reminding Test	0.845	0.033	25.65	0.85	<0.001
Visuomotor	Block Design	0.849	0.033	26.09	0.85	<0.001
Visuomotor	DSMSE Visual Spatial	0.797	0.038	17.05	0.80	<0.001
Visuomotor	Purdue Pegboard	0615	0.59	10.39	0.62	<0.001
Visuomotor	RADD Sensorimotor	0.624	0.058	10.70	0.62	<0.001
Visuomotor	VMI	0.708	0.048	14.77	0.71	<0.001

CR = critical ratio, EF = executive function, MCI = mild cognitive impairment, PID = premorbid level of ID, SE = standard error.

The largest language/EF differences between the CS and MCI-DS groups were observed on the following tests: Boston Naming, DSMSE Language, RADD-2 Expressive Language, RADD-2 Digit Span Forward, and RADD-2 Similarities (SSs = 0.76 to 0.88). The largest differences between the two groups on memory tests were on DSMSE Memory and SRT (SSs = 0.85 to 0.86). The largest differences between the two groups on visuomotor tests were on Block Design and DSMSE Visual Spatial (SSs = 0.80 to 0.85).

4. Discussion

The population with DS is unique from other populations at risk for AD due to atypical brain development, lifelong amyloid overproduction, and significant but variable cognitive impairments present prior to clinical manifestations of AD. The heterogeneity in premorbid functioning, lack of clear diagnostic criteria for this population, and use of cognitive tests that were designed for neurotypical populations have contributed to some of the difficulties in characterizing their cognitive changes during the prodromal stage of AD [19,49]. Our study of 144 older adults with DS confirmed that our extensive neuropsychological battery indeed measured cognitive functioning in at least 3 domains: language/executive functioning, memory, and visuomotor. These domains are widely investigated amongst neurotypical populations, but few studies have demonstrated that the cognitive tests used in research for people with DS measure such functions in older adults experiencing AD-related neuropathology.

Accordingly, it is important to highlight that a few of the tests in our battery that were expected to measure executive function had to be dropped from our models because of high error rates, likely associated with low comprehension of task demands. Furthermore, the tasks that were retained failed to show predicted distinctions between language and executive function skills. This underscores the need to identify or develop other tasks of executive function for this population and to evaluate their relationship(s) with tasks targeting different underlying processes—for example, Category Fluency, loaded on the language and visuomotor domains instead of with the other executive function measures, and mMMSE Fine Motor, loaded on the memory domain instead of the visuomotor domain because the task required recall of numbers and letters in the correct order. These are important reminders that cognition is multifaceted and related to multiple brain networks. Few cognitive tests solely measure one specific cognitive skill [50], and to some extent, successful performance on any test requires a certain amount of motivation, attention, comprehension, and memory. For the purposes of this paper, we only selected measures that loaded on a single domain to facilitate the interpretation of results.

4.1. Age-Related Impairments

Since aging is closely related to AD progression, we wanted to identify test outcomes that were most affected by the normal aging process and to examine whether age-related impairments differed by disease status. Aging significantly impacted performance on more than half of the cognitive tests (9 of 17) and in every domain. However, there was only one significant age x disease status interaction (RADD-2 Sensorimotor), indicating that AD progression may accelerate age-related declines on visuomotor tasks. That is, age may be a moderator, but it is not a confounder.

4.2. Premorbid Functioning

Differences in premorbid intellectual capacity did not alter the factor structure of our neuropsychological battery, but it did affect performance on all the cognitive measures. As expected, the group with mild premorbid ID consistently scored higher than the group with moderate premorbid ID. This emphasizes the importance of taking premorbid functioning into consideration in all analyses for this population due to such heterogeneity in baseline cognitive abilities. The use of a universal criterion, such as 1.5 SD below the population mean, is not appropriate for people with DS, even if the population mean is derived from other individuals with DS, because baseline IQ scores in this population can range from 2 to 4 SDs below normative levels (i.e., borderline functioning to profound ID). Nevertheless, our results provide evidence that cognitive functioning can be characterized by the same cognitive tests and domains in this population when their premorbid impairment is in the mild to moderate range.

4.3. Composites vs. Individual Test Scores

All three domain composites, which weighted all tests equally, were significantly related to MCI-DS status, but they only explained half of the total variance in test scores. In contrast, individual test scores and their regression weights explained most of the variance and had good model fit. Therefore, the cognitive features of prodromal AD in DS may be better ascertained by using re-scaled individual test scores rather than by using a domain average or sum. Additionally, the domain that was identified as being the most impacted by prodromal AD using composite scores was different than the one identified using regression weights. Composite scores indicated language/executive function skills to be the most affected, whereas individual test scores indicated the latent variable of memory to be the most affected. This accentuates the strength of estimating each domain using a factor analysis as opposed to a priori sum or mean scores.

This study demonstrates the possibility of drawing different conclusions from the same sample and outcome measures based upon the type of statistical analyses used and the factors selected for inclusion as covariates. Differences in analytical approaches may explain the conflicting findings reported between several large-scale studies on adults with DS since the neuropathological cascade caused by underlying AD should be similar for all persons with DS. For example, Cosgrave et al. [51] and Krinsky-McHale et al. [15] found memory decline to be the earliest sign of AD, whereas Ball et al. [20], Adams and Oliver [52], and Fonseca et al. [49] found executive function decline (e.g., planning, inhibitory control, working memory, and abstract thinking) to be the earlier indicator of AD progression. Still others have found language decline to be more indicative of early AD [53]. Future studies should explore the psychometric properties of their variables before deciding whether to use multiple scales or one unidimensional sum score. As McNeish and Wolf [54] illustrated, sum scoring can lead to different conclusions compared to more rigorous methods of factor estimation, and multilevel models, growth models, or multiple regression based on sum scores may be adversely affected by imprecision when summing multiple scales.

Finally, our sample of adults with DS in the pre-clinical stages of AD offers normative data on a broad array of cognitive tests that have been used among older adults in this population. It is our hope that this will assist future investigations in identifying the effects of pathological and normative aging in this population.

4.4. Limitations

Despite our best attempts to maximize participants' motivation and attention during 1.5 h of cognitive testing, including the option to administer the battery over the course of two separate visits and flexibility in the number of breaks given, we cannot rule out the possibility that test performance was influenced by these factors or language comprehension skills. Another caveat is that alternative procedures for measuring a hypothesized skill could tap multiple domains or other domains that were not included in our analyses. Hence, our factor solution may not apply to other cognitive batteries. Studies that use a different combination of tests should examine the underlying structure of their battery rather than rely upon our cognitive domain classifications. Another limitation of our study is that classification of disease status, although conducted by teams with extensive experience working with this population, is an inherently imperfect process (as it is for diagnosis of prodromal AD in elderly adults with a neurotypical developmental history). Moreover, there was some circularity in the use of certain measures of the cognitive battery to aid in consensus diagnoses, which we believe was minimized by including the results of a physical exam and a variety of informant-based measures.

Lastly, none of the participants in our larger, multisite study with severe ID successfully completed all cognitive measures. Hence, the findings in the current study should not be generalized to those individuals, a significant minority of adults with DS. This points to the need to develop and evaluate procedures specifically targeting AD clinical progression in adults with more severe lifelong disabilities.

5. Conclusions

The cognitive abilities of individuals with atypical brain development, early onset of AD pathology, and accelerated rates of amyloid accumulation may be characterized by the same cognitive domains described for neurotypical populations. However, the complex pathobiology of DS leads to both physical deficits and biochemical changes that can lead to multiple comorbid medical conditions, genetic and epigenetic variation, environmental factors, and stochastic events [14], all contributing to considerable heterogeneity in this population. Just as there are multiple cognitive phenotypes of MCI in the neurotypical population, there may also be more than one cognitive phenotype of MCI-DS.

Supplementary Materials: The following are available online at https://www.mdpi.com/article/10.3390/brainsci11091220/s1, Figure S1: Effects of aging on cognitive tests by Alzheimer's disease status.

Author Contributions: Conceptualization, C.L.H. and D.B.K.; methodology, S.J.K.-M. and W.S.; validation, S.J.K.-M. and M.B.P.; formal analysis, K.A.K. and J.R.-O.; investigation, C.L.H., S.J.K.-M. and M.B.P.; writing—original draft preparation, C.L.H.; writing—review and editing, J.R.-O., D.B.K., W.S., S.J.K.-M. and N.S.; visualization, K.A.K.; supervision, W.S.; project administration, H.D.R., F.L., N.S. and I.T.L.; funding acquisition, N.S., I.T.L. and W.S. All authors have read and agreed to the published version of the manuscript.

Funding: This research was funded by the National Institute on Aging and the Eunice Kennedy Shriver National Institute for Child Health and Human Development, grant number U01 AG051412.

Institutional Review Board Statement: The study was approved by the Institutional Review Boards of New York State Institute for Basic Research in Developmental Disabilities (protocol #7259, 04/20/2021-04/19/2022), Columbia University Irving Medical Center (protocol #AAAR0398, 02/10/2021-02/09/2022 and protocol #AAAS1197, 08/25/2021-08/24/2022), Massachusetts General Hospital (protocol #2016P000419, 02/08/2021-02/02/2022), the University of California at Irvine (protocol #2016-2690, 01/08/2021-01/07/2022), and the University of North Texas Health Science Center (protocol #2015-171, 09/02/2021-09/02/2022).

Informed Consent Statement: Informed consent was obtained from participants or their legally authorized representatives along with participant assent.

Data Availability Statement: The data presented in this study are openly available in the LONI Image and Data Archive (IDA) repository at https://ida.loni.usc.edu/login.jsp?project=ABCDS (accessed on 28 October 2020).

Acknowledgments: The authors would like to thank the adults with Down syndrome who volunteered as participants in this study for their invaluable contributions to this work, as well as their families and care providers for their generous support. We would also like to thank Deborah Pang, Eric Doran, and Courtney Jordan for their coordination of data collection.

Conflicts of Interest: The authors declare no conflict of interest. The funders had no role in the design of the study; in the collection, analyses, or interpretation of data; in the writing of the manuscript, or in the decision to publish the results.

References

1. Rafii, M.S.; Skotko, B.G.; McDonough, M.E.; Pulsifer, M.; Evans, C.; Doran, E.; Muranevici, G.; Kesslak, P.; Abushakra, S.; Lott, I.T. A Randomized, Double-Blind, Placebo-Controlled, Phase II Study of Oral ELND005 (scyllo-Inositol) in Young Adults with Down Syndrome without Dementia. *J. Alzheimer's Dis.* **2017**, *58*, 401–411. [CrossRef]

2. Koehl, L.; Harp, J.; Van Pelt, K.L.; Head, E.; Schmitt, F.A. Longitudinal assessment of dementia measures in Down syndrome. *Alzheimer's Dement.* **2020**, *12*, e12075. [CrossRef]

3. Grieco, J.; Pulsifer, M.; Seligsohn, K.; Skotko, B.; Schwartz, A. Down syndrome: Cognitive and behavioral functioning across the lifespan. *Am. J. Med. Genet. C Semin. Med. Genet.* **2015**, *169*, 135–149. [CrossRef]

4. Contestabile, A.; Benfenati, F.; Gasparini, L. Communication breaks-Down: From neurodevelopment defects to cognitive disabilities in Down syndrome. *Prog. Neurobiol.* **2010**, *91*, 1–22. [CrossRef] [PubMed]

5. Vicari, S.; Bellucci, S.; Carlesimo, G.A. Visual and spatial long-term memory: Differential pattern of impairments in Williams and Down syndromes. *Dev. Med. Child Neurol.* **2005**, *47*, 305–311. [CrossRef] [PubMed]

6. American Psychiatric Association. *Diagnostic and Statistical Manual of Mental Disorders*, 4th ed.; American Psychiatric Association: Washington, DC, USA, 1997.

7. Bondi, M.W.; Edmonds, E.C.; Jak, A.J.; Clark, L.R.; Delano-Wood, L.; McDonald, C.R.; Nation, D.A.; Libon, D.J.; Au, R.; Galasko, D.; et al. Neuropsychological criteria for mild cognitive impairment improves diagnostic precision, biomarker associations, and progression rates. *J. Alzheimer's Dis.* **2014**, *42*, 275–289. [CrossRef] [PubMed]

8. Petersen, R.C. New clinical criteria for the Alzheimer's disease spectrum. *Minn. Med.* **2012**, *95*, 42–45. [PubMed]

9. Esbensen, A.J.; Hooper, S.R.; Fidler, D.; Hartley, S.L.; Edgin, J.; D'Ardhuy, X.L.; Capone, G.; Conners, F.A.; Mervis, C.B.; Abbeduto, L.; et al. Outcome Measures for Clinical Trials in Down Syndrome. *Am. J. Intellect. Dev. Disabil.* **2017**, *122*, 247–281. [CrossRef]

10. Carr, J.; Collins, S. 50 years with Down syndrome: A longitudinal study. *J. Appl. Res. Intellect. Disabil.* **2018**, *31*, 743–750. [CrossRef]

11. Moran, J.A.; Rafii, M.S.; Keller, S.M.; Singh, B.K.; Janicki, M. The National Task Group on Intellectual Disabilities and Dementia Practices consensus recommendations for the evaluation and management of dementia in adults with intellectual disabilities. *Mayo Clin. Proc.* **2013**, *88*, 831–840. [CrossRef]

12. Basten, I.A.; Boada, R.; Taylor, H.G.; Koenig, K.; Barrionuevo, V.L.; Brandão, A.C.; Costa, A.C.S. On the Design of Broad-Based Neuropsychological Test Batteries to Assess the Cognitive Abilities of Individuals with Down Syndrome in the Context of Clinical Trials. *Brain Sci.* **2018**, *8*, 205. [CrossRef]

13. Hithersay, R.; Baksh, R.A.; Startin, C.M.; Wijeratne, P.; Hamburg, S.; Carter, B.; Strydom, A. Optimal age and outcome measures for Alzheimer's disease prevention trials in people with Down syndrome. *Alzheimer's Dement.* **2020**, *17*, 595–604. [CrossRef]

14. Cipriani, G.; Danti, S.; Carlesi, C.; Di Fiorino, M. Aging With Down Syndrome: The Dual Diagnosis: Alzheimer's Disease and Down Syndrome. *Am. J. Alzheimer's Dis. Other Dement.* **2018**, *33*, 253–262. [CrossRef]

15. Krinsky-McHale, S.J.; Devenny, D.A.; Silverman, W.P. Changes in explicit memory associated with early dementia in adults with Down's syndrome. *J. Intellect. Disabil. Res.* **2002**, *46*, 198–208. [CrossRef] [PubMed]

16. Sabbagh, M.; Edgin, J. Clinical Assessment of Cognitive Decline in Adults with Down Syndrome. *Curr. Alzheimer's Res.* **2016**, *13*, 30–34. [CrossRef] [PubMed]

17. Fortea, J.; Vilaplana, E.; Carmona-Iragui, M.; Benejam, B.; Videla, L.; Barroeta, I.; Fernández, S.; Altuna, M.; Pegueroles, J.; Montal, V.; et al. Clinical and biomarker changes of Alzheimer's disease in adults with Down syndrome: A cross-sectional study. *Lancet* **2020**, *395*, 1988–1997. [CrossRef]

18. Coppus, A.M.; Schuur, M.; Vergeer, J.; Janssens, A.C.J.; Oostra, B.A.; Verbeek, M.M.; van Duijn, C.M. Plasma β amyloid and the risk of Alzheimer's disease in Down syndrome. *Neurobiol. Aging* **2012**, *33*, 1988–1994. [CrossRef]

19. Krinsky-McHale, S.J.; Zigman, W.B.; Lee, J.H.; Schupf, N.; Pang, D.; Listwan, T.; Kovacs, C.; Silverman, W. Promising outcome measures of early Alzheimer's dementia in adults with Down syndrome. *Alzheimer's Dement. Diagn. Assess. Dis. Monit.* **2020**, *12*, e12044. [CrossRef]

20. Ball, S.L.; Holland, A.J.; Treppner, P.; Watson, P.C.; Huppert, F.A. Executive dysfunction and its association with personality and behaviour changes in the development of Alzheimer's disease in adults with Down syndrome and mild to moderate learning disabilities. *Br. J. Clin. Psychol.* **2008**, *47*, 1–29. [CrossRef]

21. Deb, S.; Hare, M.; Prior, L. Symptoms of dementia among adults with Down's syndrome: A qualitative study. *J. Intellect. Disabil. Res.* **2007**, *51*, 726–739. [CrossRef]

22. Dalton, A.J.; Mehta, P.D.; Fedor, B.L.; Patti, P.J. Cognitive changes in memory precede those in praxis in aging persons with Down Syndrome. *J. Intellect. Dev. Disabil.* **1999**, *24*, 169–187. [CrossRef]

23. Cooper, S.A.; Collacott, R.A. The effect of age on language in people with Down's syndrome. *J. Intellect. Disabil. Res.* **1995**, *39*, 197–200. [CrossRef] [PubMed]

24. Margallo-Lana, M.L.; Moore, P.B.; Kay, D.W.K.; Perry, R.H.; Reid, B.E.; Berney, T.P.; Tyrer, S.P. Fifteen-Year Follow-Up of 92 Hospitalized Adults with Down's Syndrome: Incidence of Cognitive Decline, Its Relationship to Age and Neuropathology. *J. Intellect. Disabil. Res.* **2007**, *51*, 463–477. [CrossRef] [PubMed]

25. Ball, S.L.; Holland, A.J.; Hon, J.; Huppert, F.A.; Treppner, P.; Watson, P.C. Personality and behaviour changes mark the early stages of Alzheimer's disease in adults with Down's syndrome: Findings from a prospective population-based study. *Int. J. Geriatr. Psychiatry* **2006**, *21*, 661–673. [CrossRef] [PubMed]

26. Collacott, R.A.; Cooper, S.-A. Adaptive behavior after depressive illness in Down's syndrome. *J. Nerv. Ment. Dis.* **1992**, *180*, 468–470. [CrossRef] [PubMed]

27. Lautarescu, B.A.; Holland, A.J.; Zaman, S.H. The Early Presentation of Dementia in People with Down Syndrome: A Systematic Review of Longitudinal Studies. *Neuropsychol. Rev.* **2017**, *27*, 31–45. [CrossRef]

28. Sinai, A.; Hassiotis, A.; Rantell, K.; Strydom, A. Assessing Specific Cognitive Deficits Associated with Dementia in Older Adults with Down Syndrome: Use and Validity of the Arizona Cognitive Test Battery (ACTB). *PLoS ONE* **2016**, *11*, e0153917. [CrossRef]

29. Silverman, W.; Schupf, N.; Zigman, W.; Devenny, D.; Miezejeski, C.; Schubert, R.; Ryan, R. Dementia in adults with mental retardation: Assessment at a single point in time. *Am. J. Ment. Retard.* **2004**, *109*, 111–125. [CrossRef]

30. Beery, K.E.; Buktenica, N.A. *Developmental Test of Visual Motor Integration*, 6th ed.; Pearson: Bloomington, IN, USA, 2010.

31. Wechsler, D. The Wechsler Intelligence Scale for Children. In *Technical and Interpretive Manual*, 4th ed.; The Psychological Corporation: San Antonio, TX, USA, 2003.

32. Haxby, J.V. Neuropsychological evaluation of adults with Down's syndrome: Patterns of selective impairment in non-demented old adults. *J. Ment. Defic. Res.* **1989**, *33*, 193–210. [CrossRef]

33. Kaplan, E.; Goodglass, H.; Weintraub, S. *The Boston Naming Test*; Lea & Febiger: Philadelphia, PA, USA, 1983.

34. McCarthy, D. *Manual for the McCarthy Scales of Children's Abilities*; The Psychological Corp: New York, NY, USA, 1972.

35. Stroop, J.R. Studies of interference in serial verbal reactions. *J. Exp. Psychol.* **1935**, *18*, 643–662. [CrossRef]

36. Devenny, D.A.; Zimmerli, E.J.; Kittler, P.; Krinsky-McHale, S.J. Cued recall in early-stage dementia in adults with Down's syndrome. *J. Intellect. Disabil. Res.* **2002**, *46*, 472–483. [CrossRef]

37. Folstein, M.F.; Folstein, S.E.; McHugh, P.R. Folstein Stand Mini-Mental State Exam. *J. Psychiatr. Res.* **1975**, *12*, 189–198. [CrossRef]
38. Wisniewski, K.E.; Wisniewski, H.M.; Wen, G.Y. Occurrence of neuropathological changes and dementia of Alzheimer's disease in Down's syndrome. *Ann. Neurol.* **1985**, *17*, 278–282. [CrossRef]
39. Hom, C.L.; Walsh, D.; Fernandez, G.; Tournay, A.; Touchette, P.; Lott, I.T. Cognitive assessment using the Rapid Assessment for Developmental Disabilities, Second Edition (RADD-2). *J. Intellect. Disabil. Res.* in press. **2021**. [CrossRef]
40. Parks, S. *Inside Hawaii Early Learning Profile, Birth to Three Years*; VORT Corporation: Palo Alto, CA, USA, 2006.
41. Roid, F.H.; Sampers, J.L. *Merrill-Palmer Scales of Development—Revised*; Stoelting Company: Wood Dale, IL, USA, 2004.
42. Vega, A. Use of Purdue pegboard and finger tapping performance as a rapid screening test for brain damage. *J. Clin. Psychol.* **1969**, *25*, 255–258. [CrossRef]
43. Wilson, B.; Ivani-Chalian, C.F.; Aldrich, F. *Rivermead Behavioural Memory Test for Children*; Thames Valley Test Co.: Bury St. Edmunds, UK, 1991.
44. Buschke, H. Selective reminding for analysis of memory and learning. *J. Verbal Learn. Verbal Behav.* **1973**, *12*, 543–550. [CrossRef]
45. Tinetti, M.E. Performance-oriented assessment of mobility problems in elderly patients. *J. Am. Geriatr. Soc.* **1986**, *34*, 119–126. [CrossRef] [PubMed]
46. Little, T.D. *Longitudinal Structural Equation Modeling*; The Guilford Press: New York, NY, USA, 2013.
47. Pituch, K.A.; Stevens, J.P. *Applied Multivariate Statistics for the Social Sciences*, 6th ed.; Routledge: New York, NY, USA, 2016.
48. Schumacker, R.E.; Lomax, R.G. *A Beginner's Guide to Structural Equation Modeling*, 4th ed.; Routledge: New York, NY, USA, 2016.
49. Fonseca, L.M.; Yokomizo, J.E.; Bottino, C.M.; Fuentes, D. Frontal Lobe Degeneration in Adults with Down Syndrome and Alzheimer's Disease: A Review. *Dement. Geriatr. Cogn. Disord.* **2016**, *41*, 123–136. [CrossRef]
50. Devenny, D.A.; Krinsky-McHale, S.J.; Sersen, G.; Silverman, W.P. Sequence of cognitive decline in dementia in adults with Down's syndrome. *J. Intellect. Disabil. Res.* **2000**, *44*, 654–665. [CrossRef]
51. Cosgrave, M.P.; Tyrrell, J.; McCarron, M.; Gill, M.; Lawlor, B.A. A five year follow-up study of dementia in persons with Down's syndrome: Early symptoms and patterns of deterioration. *Ir. J. Psychol. Med.* **2000**, *17*, 5–11. [CrossRef]
52. Adams, D.; Oliver, C. The relationship between acquired impairments of executive function and behaviour change in adults with Down syndrome. *J. Intellect. Disabil. Res.* **2010**, *54*, 393–405. [CrossRef] [PubMed]
53. Pulsifer, M.B.; Evans, C.L.; Hom, C.; Krinsky-McHale, S.J.; Silverman, W.; Lai, F.; Lott, I.; Schupf, N.; Wen, J.; Rosas, H.D. Language skills as a predictor of cognitive decline in adults with Down syndrome. *Alzheimer's Dement.* **2020**, *12*, e12080. [CrossRef] [PubMed]
54. McNeish, D.; Wolf, M.G. Thinking twice about sum scores. *Behav. Res. Methods* **2020**, *52*, 2287–2305. [CrossRef] [PubMed]

Article

Association between Hypothyroidism Onset and Alzheimer Disease Onset in Adults with Down Syndrome

Florence Lai [1,2,3,*], Nathaniel D. Mercaldo [1,2], Cassandra M. Wang [4], Micaela S. Hersch [5], Giovi G. Hersch [6] and Herminia Diana Rosas [1,2,3]

1 Department of Neurology, Harvard Medical School, Boston, MA 02115, USA; nmercaldo@mgh.harvard.edu (N.D.M.); rosas@helix.mgh.harvard.edu (H.D.R.)
2 Massachusetts General Hospital, Charlestown, MA 02129, USA
3 McLean Hospital, Belmont, MA 02478, USA
4 Harvard College, Harvard University, Cambridge, MA 02138, USA; cassandrawang@college.harvard.edu
5 School of Nursing, Simmons University, Boston, MA 02115, USA; micaela.hersch@simmons.edu
6 College of Arts & Sciences, Boston University, Boston, MA 02215, USA; herschgi@bu.edu
* Correspondence: flai@partners.org

Abstract: Adults with Down syndrome (DS) have an exceptionally high frequency of Alzheimer disease (AD) with a wide variability in onset, from 40 to 70 years of age. Equally prevalent in DS is hypothyroidism. In this study, we sought to quantify the relationship between the two. A total of 232 adults with DS and AD were stratified into three AD onset age groups: early (<47 years), typical (48–59), and late (>59). Among patients with available data, differences in the distributions of demographics, hypothyroidism variables (presence, age of onset), thyroid function tests, thyroid autoantibodies, and APOE genotypes were assessed (e.g., chi-squared, Mann–Whitney tests). Spearman and partial Spearman correlations and ordinal logistic regression models were constructed to quantify the association between ages of AD and hypothyroidism onset with and without covariate adjustments. We observed a positive association between the ages of AD and hypothyroidism onset after accounting for APOE-ε4 (correlation: 0.44, 0.24, 0.60; odds ratio: 1.09, 1.05–1.14). However, an early age of hypothyroidism onset and the presence of the APOE-ε4 allele were independently associated with the early age of AD onset. Similar findings were observed when accounting for other factors. Our study provides evidence for the importance of hypothyroidism and associated pathological mechanisms for risk of AD in DS.

Keywords: Down syndrome; early-onset Alzheimer disease; late-onset Alzheimer disease; hypothyroidism; thyroid autoantibodies; TSH; Free T4; APOE ε4

Citation: Lai, F.; Mercaldo, N.D.; Wang, C.M.; Hersch, M.S.; Hersch, G.G.; Rosas, H.D. Association between Hypothyroidism Onset and Alzheimer Disease Onset in Adults with Down Syndrome. *Brain Sci.* **2021**, *11*, 1223. https://doi.org/10.3390/brainsci11091223

Academic Editor: Rebecca Sims

Received: 31 July 2021
Accepted: 8 September 2021
Published: 16 September 2021

1. Introduction

Adults with Down syndrome (DS) have an especially high risk for developing Alzheimer disease (AD), with onset at least two decades earlier than in the general population [1,2]. Although there is great variation in the age of AD onset, from as early as 40 to as late as 70 or older [3,4], this variability in onset is poorly understood. The leading hypothesis for the pathogenesis of AD in DS has been attributed to overexpression of the gene for amyloid precursor protein (APP) located on the triplicated chromosome 21 [5], although factors other than amyloid likely influence the wide range of AD onset age in those with DS, just as they do in sporadic AD. Some of these factors include not only the Apolipoprotein ε4 (APOE ε4) genotype, but other genetic factors, as well as environmental or biological factors and co-existing medical conditions.

Thyroid dysfunction, including congenital, subclinical, and autoimmune thyroid conditions [6], is a very common medical co-morbidity in individuals with DS. In a large meta-analysis of over 6000 children and adults with DS, hypothyroidism was present in almost 40% [7]; similar rates (46%) of hypothyroidism were reported in another study that included only older adults [8]. Congenital hypothyroidism, believed to be primarily due to thyroid hypoplasia [9,10], is estimated to occur in a much higher frequency in children with DS than in the general population [6]. Subclinical hypothyroidism, defined as thyroid-stimulating hormone (TSH) levels above the standardized normal thyroid hormone levels, can occur in approximately one-quarter of those with DS [11]. Autoimmune thyroiditis also occurs with a high frequency in DS; autoantibodies against the thyroid such as thyroid peroxidase (TPO) antibodies are present in almost one-third of those with DS [12], with a high likelihood of conversion to overt hypothyroidism [11].

Thyroid disorders have been recognized as important risk factors in sporadic AD [13–15]. Thyroid hormones play a critical role in cognition throughout life, beginning in infancy, and deficiency has been associated with impaired cognition in memory and learning [16], similar to the cognitive impairments in AD. Indeed, hypothyroidism has been recognized as a "reversible dementia" [17], such that the standard of care in the evaluation of individuals with dementia includes screening for thyroid disorders. Both hypothyroidism and AD increase with age [18,19], and the presence of hypothyroidism has been associated with an increased risk for AD [20,21], although no direct cause–effect relationship has been established [22].

Although there is abundant evidence for the significant prevalence of hypothyroidism and AD in adults with DS [1,8,23–25], to the best of our knowledge, no prior studies have explored the possible connection between the two. Therefore, in this study, we sought to evaluate the association between a history and age of onset of hypothyroidism and the age of onset of AD in a large cohort of adults with DS.

2. Materials and Methods

IRB approval for a medical records review was obtained from Massachusetts General Hospital. We performed a retrospective study of medical records, including comprehensive medical and neurological history and cognitive assessments, of adults with Down syndrome who had been followed prospectively on an annual basis in a neurology DS subspecialty clinic. A total of 232 patients were identified who had been diagnosed with possible or probable AD, based on the criteria developed by the AAMR-IASSID Working Group for the Establishment of Criteria for the Diagnosis of Dementia in Individuals with Developmental Disability [26]. Patients were classified into two groups of premorbid level of intellectual disability (LID) based on IQ scores or functional ability: (1) mild/moderate LID: IQ between 40 and 70, ability to perform most activities of daily living (ADL), and reasonable language skills; (2) severe/profound LID: IQ < 40, needing at least some assistance in ADLs, with limited language skills.

2.1. Clinical Assessments

Patients underwent annual clinical evaluations and cognitive assessments, including tests standardized for use in individuals with DS (Test of Severe Impairment [27], Verbal Fluency Test [28] and the Dementia Questionnaire for People with Learning Disabilities [26]). Dementia status was determined by a neurologist with experience in diagnosing AD in the DS population. Blood collected at the time of clinical diagnosis of AD included thyroid function tests (TSH and Free T4), Vitamin B12 levels, and APOE genotyping; in a subset of the cohort, blood was sent for the evaluation of thyroid autoantibodies including TPO antibodies and/or thyroglobulin (Tg).

Age of onset of AD was determined based on at least a one-year progressive decline in two or more cognitive domains. Early onset was defined as having an AD onset greater than one standard deviation (SD) below the mean age of AD onset for the cohort, and late onset was defined as greater than one SD above the mean age of AD onset for this group. In our cohort, the mean (SD) age of AD onset was 53 (6). Therefore, early AD onset refers to onset before the age 47, typical AD onset as occurring between ages 47 and 59, and late AD onset as occurring after the age of 59.

Age of hypothyroidism onset, which was obtained from caregivers and available for 111 of the 152 patients (73%) with a reported history of hypothyroidism, was the age at which thyroid supplementation was first started based on abnormal thyroid function tests.

2.2. Statistical Analyses

Descriptive summaries were computed by age of AD diagnosis (early, typical, or late). Continuous variables were summarized using either the mean/standard deviation (SD) or using the median and interquartile range (IQR, 25th–75th percentile). If missing data were observed, the frequency of non-missing variable responses was augmented to the summaries of the continuous variables (e.g., mean (SD); n or median (IQR); n). Categorical variables were summarized as percentages and frequencies of non-missing responses. Using a complete-case analysis, preliminary differences in the distributions of categorical and continuous variables by diagnosis were assessed using either the chi-squared test/Fisher's exact test or the Mann–Whitney test, respectively. The Spearman correlation coefficient and its 95% confidence interval (CI) were computed to summarize the monotonic relationship between the ages of hypothyroidism and AD onset.

Partial Spearman correlation coefficients were also computed between ages of hypothyroidism onset and age of AD onset, while separately accounting for each demographic variable or co-varying medical conditions of interest (sex, APOE ε4 status, body mass index (BMI), history of vitamin B12 deficiency, history of obstructive sleep apnea (OSA), and level of intellectual disability). An exploratory series of proportional odds logistic regression models were constructed to quantify the association between the age of AD onset and age of hypothyroidism onset while separately accounting for each demographic or covarying medical condition and their interactions. Wald tests were performed to assess model complexity (non-linear terms, interaction effects). Parameter estimates, 95% CI, and *p*-values were computed to summarize the final regression models.

3. Results

3.1. Demographics

Demographics for the cohort and for each diagnostic group are provided in Table 1. The cohort included 36 individuals with early AD onset, 160 with a typical age of AD onset and 36 with late AD onset. Differences in the distribution of sex and the level of intellectual disability were not observed across the three age of AD onset groups ($p = 0.776$ and $p = 0.265$, respectively). There was a significantly higher frequency of patients carrying an APOE ε4 allele (ε3/4 and ε4/4) in the early AD onset group than in the later AD onset groups ($p = 0.040$).

A history of hypothyroidism was present in 58.3% of the early, 65.6% of the typical and 72.2% of the late AD onset cases; there was no significant difference in the frequency of hypothyroidism amongst groups ($p = 0.463$). There was no significant difference in BMI at the time of AD diagnosis ($p = 0.970$). There was, however, a higher prevalence of a diagnosis of OSA in the early AD onset group ($p = 0.048$).

Given previous reports of the known relationship between thyroid dysfunction and cognitive impairment [16,29], we sought to confirm that individuals were euthyroid at the time of AD diagnosis (Table 2). Among patients with a measured TSH value, we were unable to detect differences in the distribution of these values by age of AD onset (continuous: $p = 0.616$, categorized TSH: $p = 0.310$). Approximately 3% of the early, 7% of the typical, and 9% of the late AD onset groups had TSH levels less than 0.34 uIU/mL, suggesting a possible hyperthyroid state. Approximately 3% of the early, 15% of the typical, and 15% of the late AD onset patients had TSH levels greater than 5 uIU/mL, suggesting that they may have been inadequately treated at the time of AD diagnosis. Differences in the distributions of Free T4 levels were not detected amongst the three groups ($p = 0.277$).

Table 1. Demographic summaries by age of AD onset group.

	Early	Typical	Late	
	N = 36	**N = 160**	**N = 36**	***p***
AD onset (years)				<0.001
Mean (SD)	43.81 (1.92)	52.92 (3.72)	62.19 (2.30)	
Sex, % (n)				0.776
Male	61.1 (22)	55.0 (88)	58.3 (21)	
Female	38.9 (14)	45.0 (72)	41.7 (15)	
Level of Intellectual Disability, % (n)				0.265
Mild/Moderate	72.2 (26)	62.8 (98)	53.1 (17)	
Profound/Severe	27.8 (10)	37.2 (58)	46.9 (15)	
APOE ε4 allele, % (n)				0.04
Absent	62.1 (18)	77.0 (107)	90.0 (27)	
Present (3/4,4/4)	37.9 (11)	23.0 (32)	10.0 (3)	
BMI (kg/m^2)				0.97
Mean (SD); n	30.19 (5.06); 26	29.93 (6.34); 123	30.14 (4.67); 28	
History of hypothyroidism, % (n)				0.463
No	41.7 (15)	34.4 (55)	27.8 (10)	
Yes	58.3 (21)	65.6 (105)	72.2 (26)	
Hypothyroidism onset (years)				0.003
Mean (SD); n	34.1 (10.86); 17	42.2 (9.57); 78	45.7 (11.40); 16	
History of obstructive sleep apnea, % (n)				0.048
Absent	61.1 (22)	75.4 (120)	86.1 (31)	
Present	38.9 (14)	24.5 (39)	13.9 (5)	
History of vitamin B12 deficiency, % (n)				0.993
Absent	80.6 (29)	79.9 (127)	80.6 (29)	
Present	19.4 (7)	20.1 (32)	19.4 (7)	

Table 2. Thyroid function tests: summaries by age of AD onset group.

	Early	Typical	Late	
	N = 36	**N = 160**	**N = 36**	***p***
TSH (ulU/mL, continuous)				0.616
Median (IQR); n	2.46 (1.78, 3.04); 33	2.04 (1.23, 3.63); 148	2.10 (1.18, 3.37); 34	
TSH (ulU/mL, categorical), % (n)				0.31
0.00–0.33	3.0 (1)	6.8 (10)	8.8 (3)	
0.34–5.00	93.9 (31)	78.4 (116)	76.5 (26)	
>5.00	3.0 (1)	14.9 (22)	14.7 (5)	
Free T4 (ng/dL, continuous)				0.277
Median (IQR); n	0.90 (0.83, 1.08); 18	1.00 (0.90, 1.20); 95	0.95 (0.90, 1.30); 28	
Free T4 (ng/dL, categorical), % (n)				0.832
0.00–0.89	27.8 (5)	20.0 (19)	21.4 (6)	
0.90–1.90	72.2 (13)	77.9 (74)	78.6 (22)	
>1.90	0.0 (0)	2.1 (2)	0.0 (0)	

3.2. Association between the Age of Onset of AD and Age of Onset of Hypothyroidism

Among individuals with a history of hypothyroidism and a reported age of hypothyroidism onset, we observed a significant difference in the age of hypothyroidism by the age of AD onset (p = 0.003; Table 1). More specifically, individuals belonging to the early AD onset group had a significantly earlier age of a diagnosis of hypothyroidism, followed by the typical onset, and then by the late AD onset group. Importantly, at the time of evaluation for AD, the majority of the cohort was euthyroid; there was no significant difference amongst the cohorts with respect to TSH or Free T_4 blood concentrations.

We subsequently evaluated the association between the age of hypothyroidism onset and age of AD onset. The Spearman correlation between the age of hypothyroidism onset and age of AD onset was 0.43 (95% CI: 0.27, 0.57) ($p < 0.001$). When assuming a linear representation of age of hypothyroidism ($p_{anova} > 0.05$), we observed that for each year increase in the age of hypothyroidism onset, the unadjusted odds of having a later age of AD onset increased by a factor of 1.09 (95% CI: 1.05–1.12). Thus, among patients with hypothyroidism, those who developed it earlier appeared to have an earlier age of AD onset.

Similar findings were observed after accounting for each demographic variable and co-varying medical condition of interest (history of vitamin B12 deficiency: 0.44 (0.25, 0.59); history of OSA: 0.45 (0.26, 0.60); BMI: 0.44 (0.22, 0.62); APOE ε4 status: 0.44 (0.24, 0.60), sex: 0.43 (0.24, 0.58); level of intellectual disability: 0.46 (0.28, 0.61); all $p < 0.001$). There was insufficient evidence to conclude that a non-linear coding of age of hypothyroidism onset and the interaction between age of hypothyroidism onset and each demographic variable improved the model fit of age of AD onset (all $p_{anova} > 0.05$). Thus, we observed that for each year increase in the age of hypothyroidism onset, the adjusted odds of having a later age of AD onset increased by a factor of: 1.09 (95% CI: 1.05–1.12) after adjusting for history of vitamin B12 deficiency, 1.08 (1.05, 1.12) after adjusting for sex, 1.09 (1.05, 1.13) after adjusting for history of OSA, 1.10 (1.05, 1.14) after adjusting for BMI, 1.10 (1.06, 1.14) after adjusting for level of intellectual disability, and 1.09 (1.05, 1.14) after adjusting for APOE ε4 status.

3.3. Thyroid Autoantibodies

To evaluate the potential contribution of autoimmune thyroiditis to the risk of developing AD early, we evaluated TPO antibody levels; these were available for only approximately 41% (96/232) of the entire cohort (Table 3). Among those with a recorded TPO value, we were unable to detect differences in the distribution of TPO values by the age of AD onset ($p = 0.591$). The frequency of an elevated TPO, defined as a level above 9 IU/mL, was present in approximately 33% of the early, 40% of the typical, and 32% of the late AD onset group ($p = 0.731$).

Table 3. Thyroid antibody summaries by age of AD onset group.

	Early	Typical	Late	p
	N = 36	N = 160	N = 36	
TPO (IU/mL, continuous)				0.591
Median (IQR); n	1.30 (1.10, 16.50); 9	3.75 (1.40, 24.62); 64	4.01 (1.53, 10.30); 22	
TPO (IU/mL, categorical), % (n)				0.731
0.00–9.00	66.7 (6)	59.4 (38)	68.2 (15)	
>9.00	33.3 (3)	40.6 (26)	31.8 (7)	
Tg (IU/mL, continuous)				0.533
Median (IQR); n	2.65 (1.80, 3.88); 8	1.80 (1.80, 2.45); 50	1.80 (1.80, 1.80); 16	
Tg (IU/mL, categorical), % (n)				0.701
0.00–4.00	75.0 (6)	86.0 (43)	81.2 (13)	
>4.00	25.0 (2)	14.0 (7)	18.8 (3)	

Thyroglobulin (Tg) antibody levels were available for approximately 32% (74/232) of the cohort. Among those with a recorded Tg value, values tended to be less than 4 IU/mL, but we were unable to detect differences in the distribution of these values by the age of AD onset ($p = 0.533$).

4. Discussion

It is well known that both hypothyroidism and AD occur in high frequency in those with DS; however, our study is the first to explore the potential relationship between the age of onset of hypothyroidism and age of onset of AD in DS. Although an earlier age of AD onset in the DS population has been reported to occur, on average, 20 years earlier than in the neurotypical population, it has generally been ascribed to the triplication of APP on chromosome 21. However, our data suggest that the presence of hypothyroidism early in life may also provide its own contribution to risk for AD in DS. This was true even when other co-variates and other medical co-morbidities known to independently contribute to cognitive dysfunction were taken into consideration. Earlier onset of hypothyroidism within the DS population was associated with an even greater risk for the earlier onset of AD. Although the biological mechanisms that might explain this are as yet unknown, one possibility is that thyroid hormone itself may impact the expression of amyloid precursor protein, such that reduced thyroid hormone levels increase APP expression and elevations of pathogenic amyloid [30]. This effect may be even more significant in those with DS who have a much higher amyloid burden.

As expected, the presence of an APOE ε4 allele was associated with an earlier onset of AD in DS, as has been reported by others [31]; however, we did not detect an interaction between the presence of the APOE ε4 allele and the relationship between the age of onset of AD and age of hypothyroidism onset. Our findings suggest that APOE ε4 and early age of hypothyroidism may each contribute independently to the risk for early AD onset in the DS population.

In a subset of our cohort for whom results for serum thyroid autoantibodies were available, approximately 40% had thyroid autoantibodies as evidence of autoimmune thyroiditis, consistent with what has been reported previously in DS [32]. A similar frequency of thyroid autoantibodies was present in each of the three age of AD onset groups, suggesting that the presence of autoantibodies did not confer added risk for the development of early AD in the DS population. However, we did not have sufficient sample size to evaluate the potential interaction with the presence of thyroid autoantibodies. Nevertheless, given the comparable distribution and similar blood levels across the three groups, it appears unlikely that thyroid autoantibodies confer added risk for early AD onset in the DS population. Future studies are needed to fully evaluate the potential contribution of autoimmune thyroiditis to AD risk in DS.

It is important to point out that at the time of an AD diagnosis, a majority of patients in each of the cohorts had TSH levels within the normal ranges, confirming that the majority were euthyroid at the time of AD diagnosis, and that there were no significant differences amongst TSH levels across the groups. Similarly, the overall frequency of hypothyroidism in our sample was consistent with the findings in a large meta-analysis of adults with DS [8], indicating that our cohort is representative of the DS population at large. A diagnosis of hypothyroidism appeared to occur, on average, more than a decade prior to a diagnosis of AD (as defined as the difference between the age of AD onset and age of hypothyroidism onset), irrespective of age of AD onset. Alterations in thyroid function have been associated with a higher risk of developing AD later in life in the neurotypical population [33], suggesting that complex interactions between thyroid hormone, thyroid function and risk for AD may be particularly salient in the DS population.

5. Limitations

The primary limitations of this study include: (1) the retrospective nature of the study design; (2) the possible measurement error associated with ages of AD onset and hypothyroidism onset; and (3) the degree of missingness especially involving the thyroid autoantibody panels. The validity of these results assumes that the covariate missingness mechanism is completely random. Nevertheless, the results of this study need to be studied further in larger (and ideally prospectively collected) datasets.

6. Conclusions

In conclusion, our study suggests that the early onset of hypothyroidism in DS is significantly associated with an early onset age of AD, and that it is independent of APOE ε4 allele status, BMI, vitamin B12 status, or presence of OSA. This emphasizes the importance of early testing for TSH, thyroid hormone and thyroid autoantibodies, and treatment with thyroid replacement as needed. Future studies are needed to determine the mechanisms by which a history of hypothyroidism affects AD risk and onset, including its relationship with other genetic, inflammatory or metabolomic alterations present in adults with DS.

Author Contributions: Conceptualization: F.L., H.D.R.; methodology: N.D.M., F.L., H.D.R., C.M.W., M.S.H., G.G.H.; formal analysis: N.D.M.; data curation: C.M.W., M.S.H., G.G.H.; writing—review and editing, all authors. All authors have read and approved the submitted version and agree to be personally accountable for the author's own contributions and for ensuring that questions related to the accuracy or integrity of any part of the work, even ones in which the author was not personally involved, are appropriately investigated, resolved, and documented in the literature. All authors have read and agreed to the published version of the manuscript.

Funding: This research was funded by the Libbi Thomas Foundation.

Institutional Review Board Statement: The study was conducted according to the guidelines of the Declaration of Helsinki and approved by the Institutional Review Board of Massachusetts General Hospital (Protocol # 2016P00198).

Informed Consent Statement: Patient consent was waived due to the retrospective nature of the study.

Data Availability Statement: Data are available for review by request to corresponding author.

Acknowledgments: The authors are grateful to the many adults with Down syndrome who entrusted themselves to our clinical care and to their families and caregivers who supported them. We thank Courtney Jordan and Kelsey Shelofsky for their assistance.

Conflicts of Interest: The authors declare no conflict of interest. The funding foundation had no role in the design of the study, in the collection of the data, in the statistical analysis, in the interpretation of the data or in the writing of the manuscript.

References

1. Lai, F.; Williams, R.S. A Prospective Study of Alzheimer Disease in Down Syndrome. *Arch. Neurol.* **1989**, *46*, 849–853. [CrossRef] [PubMed]
2. McCarron, M.; McCallion, P.; Reilly, E.; Dunne, P.; Carroll, R.; Mulryan, N. A prospective 20-year longitudinal follow-up of dementia in persons with Down syndrome. *J. Intellect. Disabil. Res.* **2017**, *61*, 843–852. [CrossRef]
3. Davidson, Y.S.; Robinson, A.; Prasher, V.P.; Mann, D.M.A. The age of onset and evolution of Braak tangle stage and Thal amyloid pathology of Alzheimer's disease in in-dividuals with Down syndrome. *Acta Neuropathol. Commun.* **2018**, *6*, 56. [CrossRef] [PubMed]
4. Schupf, N.; Lee, J.H.; Pang, D.; Zigman, W.; Tycko, B.; Krinsky-McHale, S.; Silverman, W. Epidemiology of estrogen and dementia in women with Down syndrome. *Free Radic. Biol. Med.* **2018**, *114*, 62–68. [CrossRef]
5. Mann, D.M. The pathological association between down syndrome and Alzheimer disease. *Mech. Ageing Dev.* **1988**, *43*, 99–136. [CrossRef]
6. Amr, N.H. Thyroid disorders in subjects with Down syndrome: An update. *Acta Biomed.* **2018**, *89*, 132–139. [CrossRef]
7. Chicoine, B.; Rivelli, A.; Fitzpatrick, V.; Chicoine, L.; Jia, G.; Rzhetsky, A. Prevalence of Common Disease Conditions in a Large Cohort of Individuals With Down Syndrome in the United States. *J. Patient-Cent. Res. Rev.* **2021**, *8*, 86–97. [CrossRef]
8. Bayen, E.; Possin, K.L.; Chen, Y.; de Langavant, L.C.; Yaffe, K. Prevalence of Aging, Dementia, and Multimorbidity in Older Adults with Down Syndrome. *JAMA Neurol.* **2018**, *75*, 1399–1406. [CrossRef]
9. Kariyawasam, D.; Luton, D.; Polak, M. Down Syndrome and Nonautoimmune Hypothyroidisms in Neonates and Infants. *Horm. Res. Paediatr.* **2015**, *83*, 126–131. [CrossRef]
10. Ayşe, N.C.; Ayla, G.; Metin, Y.; Cebeci, A.N.; Güven, A.; Yıldız, M. Profile of Hypothyroidism in Down's Syndrome. *J. Clin. Res. Pediatr. Endocrinol.* **2013**, *5*, 116–120. [CrossRef] [PubMed]
11. King, K.; O'Gorman, C.; Gallagher, S. Thyroid dysfunction in children with Down syndrome: A literature review. *Ir. J. Med. Sci.* **2014**, *183*, 1–6. [CrossRef]

12. Nada, A.; Ashraf, T.S.; Maya, I.; Ahmed, K.; Vincenzo, D.S.; Alaaraj, N.; Soliman, A.T.; Itani, M.; Khalil, A.; De Sanctis, V. Prevalence of thyroid dysfunctions in infants and children with Down Syndrome (DS) and the effect of thyroxine treatment on linear growth and weight gain in treated subjects versus DS subjects with normal thyroid function: A controlled study. *Acta Biomed.* **2019**, *90*, 36–42.

13. Kalmijn, S.; Mehta, K.M.; Pols, H.A.P.; Hofman, A.; Drexhage, H.A.; Breteler, M.M. Subclinical hyperthyroidism and the risk of dementia. The Rotterdam study. *Clin. Endocrinol.* **2000**, *53*, 733–737. [CrossRef]

14. Tan, Z.S.; Vasan, R.S. Thyroid function and Alzheimer's disease. *J. Alzheimers Dis.* **2009**, *16*, 503–507. [CrossRef]

15. Chaker, L.; Wolters, F.J.; Bos, D.; Korevaar, T.I.; Hofman, A.; van der Lugt, A.; Koudstaal, P.J.; Franco, O.H.; Dehghan, A.; Vernooij, M.W.; et al. Thyroid function and the risk of dementia: The Rotterdam Study. *Neurology* **2016**, *87*, 1688–1695. [CrossRef]

16. Bavarsad, K.; Hosseini, M.; Hadjzadeh, M.A.; Sahebkar, A. The effects of thyroid hormones on memory impairment and Alzheimer's disease. *J. Cell. Physiol.* **2019**, *234*, 14633–14640. [CrossRef]

17. Cordes, J.; Cano, J.; Haupt, M. Reversible dementia in hypothyroidism. *Nervenarzt* **2000**, *71*, 588–590. [CrossRef] [PubMed]

18. Trevisan, K.; Cristina-Pereira, R.; Silva-Amaral, D.; Aversi-Ferreira, T.A. Theories of Aging and the Prevalence of Alzheimer's Disease. *BioMed Res. Int.* **2019**, *2019*, 9171424. [CrossRef] [PubMed]

19. Canaris, G.J.; Manowitz, N.R.; Mayor, G.; Ridgway, E.C. The Colorado Thyroid Disease Prevalence Study. *Arch. Intern. Med.* **2000**, *160*, 526–534. [CrossRef] [PubMed]

20. Ganguli, M.; Burmeister, L.; Seaberg, E.C.; Belle, S.; DeKosky, S. Association between dementia and elevated TSH: A community-based study. *Biol. Psychiatry* **1996**, *40*, 714–725. [CrossRef]

21. Thvilum, M.; Brandt, F.; LillevangJohansen, M.; Folkestad, L.; Brix, T.H.; Hegedüs, L. Increased risk of dementia in hypothyroidism: A Danish nationwide registerbased study. *Clin. Endocrinol.* **2021**, *94*, 1017–1024. [CrossRef] [PubMed]

22. Figueroa, P.B.S.; Ferreira, A.F.F.; Britto, L.R.; Doussoulin, A.P.; da Silva Torrão, A. Association between thyroid function and Alzheimer's disease: A systematic review. *Metab. Brain Dis.* **2021**, *36*, 1523–1543. [CrossRef] [PubMed]

23. Percy, M.E.; Potyomkina, Z.; Dalton, A.J.; Fedor, B.; Mehta, P.; Andrews, D.F.; Mazzulli, T.; Murk, L.; Warren, A.C.; Wallace, R.A.; et al. Relation between apolipoprotein E genotype, hepatitis b virus status, and thyroid status in a sample of older persons with down syndrome. *Am. J. Med. Genet.* **2003**, *120A*, 191–198. [CrossRef] [PubMed]

24. Lott, I.T.; Dierssen, M. Cognitive deficits and associated neurological complications in individuals with Down's syndrome. *Lancet Neurol.* **2010**, *9*, 623–633. [CrossRef]

25. Percy, M.E.; Dalton, A.J.; Markovic, V.D.; McLachlan, D.R.C.; Gera, E.; Hummel, J.T.; Rusk, A.C.M.; Somerville, M.J.; Andrews, D.F.; Walfish, P.G. Autoimmune thyroiditis associated with mild "subclinical" hypothyroidism in adults with down syndrome: A comparison of patients with and without manifestations of Alzheimer disease. *Am. J. Med. Genet.* **1990**, *36*, 148–154. [CrossRef]

26. Eurlings, H.A. *Dementia Questionnaire for Persons with Learning Disabilities (DLD)*; Pearson Assessment: London, UK, 2006.

27. Albert, M.S. Parallels between Down syndrome dementia and Alzheimer's disease. *Prog. Clin. Biol. Res.* **1992**, *379*, 77–102.

28. McCarthy, D. *The McCarthy Scales of Children's Abilities*; The Psychological Corporation: New York, NY, USA, 1972.

29. Winkler, A.; Weimar, C.; Jöckel, K.-H.; Erbel, R.; Dragano, N.; Broecker-Preuss, M.; Moebus, S.; Führer-Sakel, D.; Dlugaj, M. Thyroid-Stimulating Hormone and Mild Cognitive Impairment: Results of the Heinz Nixdorf Recall Study. *J. Alzheimer's Dis.* **2015**, *49*, 797–807. [CrossRef]

30. O'Barr, S.A.; Oh, J.S.; Ma, C.; Brent, G.A.; Schultz, J.J. Thyroid hormone regulates endogenous amyloid-beta precursor protein gene expression and processing in both in vitro and in vivo models. *Thyroid* **2006**, *16*, 1207–1213. [CrossRef]

31. Schupf, N.; Kapell, D.; Roudrigez, A.; Tycko, B.; Mayeux, R. Earlier onset of Alzheimer's disease in men with Down syndrome. *Neurology* **1998**, *50*, 991–995. [CrossRef]

32. Ivarsson, S.-A.; Ericsson, U.-B.; Gustafsson, J.; Forslund, M.; Vegfors, P.; Annerén, G. The impact of thyroid autoimmunity in children and adolescents with Down syndrome. *Acta Paediatr.* **1997**, *86*, 1065–1067. [CrossRef] [PubMed]

33. Kim, B.; Moon, S.W. Association between Thyroid Hormones, Apolipoprotein E, and Cognitive Function among Cognitive-ly-Normal Elderly Dwellers. *Psychiatry Investig.* **2020**, *17*, 1006–1012. [CrossRef] [PubMed]

MDPI
St. Alban-Anlage 66
4052 Basel
Switzerland
Tel. +41 61 683 77 34
Fax +41 61 302 89 18
www.mdpi.com

Brain Sciences Editorial Office
E-mail: brainsci@mdpi.com
www.mdpi.com/journal/brainsci

www.ingramcontent.com/pod-product-compliance
Lightning Source LLC
LaVergne TN
LVHW070909170726
843515LV00005B/741